Proceedings in Life Sciences

Fish Diseases

Third COPRAQ-Session

Edited by W. Ahne

With 120 Figures

Springer-Verlag
Berlin Heidelberg New York 1980

Dr. Winfried Ahne
Institut für Zoologie und Hydrobiologie
Universität München
Fachbereich Tiermedizin
Kaulbachstraße 37
8000 München 22, FRG

ISBN 3-540-10406-2 Springer-Verlag Berlin Heidelberg New York
ISBN 0-387-10406-2 Springer-Verlag New York Heidelberg Berlin

Library of Congress Cataloging in Publication Data. Main entry under title: Fish diseases. (Proceedings in life sciences) Bibliography: p. Includes index. 1. Fishes–Diseases–Congresses. I. Ahne, W., 1939–. II. Title: COPRAQ-session. SH171.F563 639.9'77 80-25293.

This work is subject to copyright. All rights are reserved, whether the whole or part of the material is concerned, specifically those of translation, reprinting, re-use of illustrations, broadcasting, reproduction by photocopying machine or similar means, and storage in data banks. Under § 54 of the German Copyright Law, where copies are made for other than private use, a fee is payable to "Verwertungsgesellschaft Wort", Munich.

© by Springer-Verlag Berlin Heidelberg 1980.
Printed in Germany.

The use of registered names, trademarks, etc. in this publication does not imply, even in the absence of a specific statement, that such names are exempt from the relevant protective laws and regulations and therefore free for general use.

Offsetprinting and bookbinding: Brühlsche Universitätsdruckerei, Giessen

2131/3130-543210

Preface

This book records contributions presented at the Third Session of the EIFAC (European Inland Fisheries Advisory Commission of the Food and Agriculture Organization of the United Nations) and OIE (Office International des Epizooties) Cooperative Programme of Research on Aquaculture (COPRAQ) Fish Diseases. The session was held at the Institute for Zoology and Hydrobiology, Veterinary Faculty, Munich, on the 23-26 October 1979.

Sponsorship of the session was by the Federal Ministry of Food Agriculture and Forest of the Federal Republic of Germany and by the Deutsche Forschungsgemeinschaft (DFG).

The EIFAC was represented by Prof. Dr. N. Fijan, Zagreb, Yugoslavia, and the OIE was represented by Prof. Dr. P. Ghittino, Turin, Italy and Dr. P. de Kinkelin, Thiverval-Grignon, France.

The Third COPRAQ Session brought to Munich an array of scientists from all over the world. Over 100 participants from more than 20 countries (most were from Europe, but Australia, Canada, Egypt, Israel, Kuwait, Nepal, and USA were also represented) discussed the major problems affecting the health of cultured fish. 30 papers for the slide sections and 13 papers for the poster section were submitted. Topics included were viral, bacterial, and parasitic diseases of fish, as well as emerging problems and approaches in aquaculture.

The papers presented at the Third COPRAQ Session pointed out that in contrast to the First COPRAQ Session (held in 1975 in Zagreb, Yugoslavia) and to the Second COPRAQ Session (held in 1977 in Brest, France) where viral fish diseases were predominant, bacterial diseases such as vibriosis, furunculosis and carp erythrodermatitis are becoming more and more important. In some cases of fish diseases immunization has been shown to be an effective method of prevention. However, control and prevention of diseases are essential for the health of animals in aquaculture.

I should like to extend my grateful thanks to all of those who actively took part in the Third COPRAQ-Session. Appreciation is also expressed to all those who helped me to organize the Session, especially to my wife Rita and to my assistants Mrs. Weiler and Miss Held. To the Springer-Verlag my thanks are due for patient help during the editing of this volume.

October, 1980   W. Ahne

Contents

Viral Diseases
Chairman: P. DE KINKELIN, France

Egtved Virus: The Susceptibility of Brown Trout and Rainbow Trout to Eight Virus Isolates and the Significance of the Findings for the VHS Control
P.E.V. JØRGENSEN (With 2 Figures) 3

Isolation of VHS Virus from Pike Fry *(Esox lucius)* with Hemorrhagic Symptoms
W. MEIER and P.E.V. JØRGENSEN (With 13 Figures) 8

Some Aspects of Trout Gill Structure in Relation to Egtved Virus Infection and Defence Mechanisms
S. CHILMONCZYK (With 6 Figures) 18

Experimental Infection of Susceptible Carp Fingerlings with Spring Viremia of Carp Virus, Under Wintering Environmental Conditions
A.M. BAUDOUY, M. DANTON, and G. MERLE (With 2 Figures) . 23

Spring Viremia of Carp (SVC): Studies on Immunization of Carp
W. AHNE 28

Studies on Immunization of Trout Against IPN
B.J. HILL, M. DORSON, and P.F. DIXON (With 2 Figures) 29

Bacterial Diseases
Chairman: A.L.S. MUNRO, Scotland

Experimental Vibrosis in the Eel *(Anguilla anguilla)*
H. CHART and C.B. MUNN (With 4 Figures) 39

Laboratory and Field Observations on Antivibriosis Vaccines
T.P.T. EVELYN and J.E. KETCHESON 45

Recent Experience of Field Vaccination Trials Against Vibriosis
in Rainbow Trout *(Salmo gairdneri)*
T. HÅSTEIN, F. HALLINGSTAD, T. REFSTI, and S.O. ROALD
(With 3 Figures) 53

Some Results of Vaccination Against Vibriosis in Brittany
F. BAUDIN LAURENCIN and J. TANGTRONGPIROS
(With 1 Figure) 60

Production and Properties of a Haemolytic Toxin by *Vibrio anguillarum*
C.B. MUNN (With 4 Figures) 69

Observations on Vibriosis in Cultured Flatfish
R.H. RICHARDS (With 1 Figure) 75

Experimental and Naturally Occurring Furunculosis in Various
Fish Species: a Comparative Study
D. BUCKE 82

Seasonal Occurrence of *Aeromonas salmonicida* Carrers
N.J. JENSEN and J.L. LARSEN 87

Examination in Resistance Tests of Some Strains of the
Aeromonas hydrophila punctata Group Isolated from Carp
W. NEUMANN and W. PLÖGER 90

Development of Bacteria in Fish and in Water During a
Standardized Experimental Infection of Rainbow Trout
(Salmo gairdneri) with *Aeromonas salmonicida*
C. MICHEL 94

Studies on an Ichthyotoxic Material Produced Extracellularly
by the Furunculosis Bacterium *Aeromonas salmonicida*
A.L.S. MUNRO, T.S. HASTINGS, A.E. ELLIS, and
J. LIVERSIDGE (With 3 Figures) 98

Studies on Vaccination of Atlantic Salmon Against Furunculosis
R. PALMER and P.R. SMITH (With 1 Figure) 107

Further Studies on Furunculosis Vaccination
P.D. SMITH, D.H. MCCARTHY, and W.D. PATERSON 113

Pathogenesis of Carp Erythrodermatitis (CE): Role of Bacterial
Endo- and Exotoxin
J.M.A. POL, R. BOOTSMA, and J.M. v.d. BERG-BLOMMAERT
(With 1 Figure) 120

Examination of the CE Agent
G. CSABA, B. KÖRMENDY, and L. BEKESI 126

Some Aspects of the Histopathology of Carp Erythrodermatitis (CE)
E.K. GAYER, L. BEKESI, and G. CSABA (With 10 Figures) ... 127

Erythrodermatitis of Carp: Studies of the Mode of Infection
D. SCHULZ (With 6 Figures) 137

Emerging Problems and Approaches
Chairman: P. GHITTINO, Italy

Characterization of the Causal Agents of Bacterial Kidney Disease
B. AUSTIN (With 3 Figures) 147

Infection with an Acinetobacter-like Bacterium in Atlantic
Salmon *(Salmo salar)* Broodfish
S.O. ROALD and T. HASTEIN 154

"Sphaerosporosis", a New Kidney Disease of the Common Carp
K. MOLNÁR (With 4 Figures) 157

Disease in Farmed Juvenile Atlantic Salmon Caused by
Dermocystidium sp.
A.H. MC VICAR and R. WOOTTEN (With 11 Figures) 165

The Causative Agent of Proliferative Kidney Disease May Be
a Member of the Haplosporidia
C. SEAGRAVE, D. BUCKE, and D. ALDERMAN
(With 6 Figures) 174

Poster Section

Preparation of Salmonid White Blood Cells for Virological Studies
P.-J. ENZMANN 185

Use of Immunoperoxidase Technique for Detection of Fish
Virus Antigens
M. FAISAL and W. AHNE (With 2 Figures) 186

Nephrocalcinosis of Rainbow Trout (*Salmo gairdneri* Richardson)
in Freshwater; a Survey of Affected Farms
J.G. HARRISON 193

The Increase of Nephrocalcinosis (NC) in Rainbow Trout
in Intensive Aquaculture
H.-J. SCHLOTFELDT (With 5 Figures) 198

Trial Vaccination of Rainbow Trout Against *Aeromonas liquefaciens*
ACUIGRUP 206

Flavobacteriosis in Coho Salmon *(Oncorhynchus kisutch)*
ACUIGRUP (With 1 Figure) 212

Hemagglutination Properties of *Aeromonas*
T.J. TRUST, I.D. COURTICE, and H.M. ATKINSON
(With 1 Figure) 218

Bacterial Stress-Caused Infections of Silver Carp and
Sarotherodon aureus in Fish Ponds and Their Control
S. SARIG and I. BEJERANAO 224

Physiological and Morphological Effects of Social Stress on the
Eel, *Anguilla anguilla* L.
GABRIELE PETERS, H. DELVENTHAL, and H. KLINGER . . . 225

Epitheliocystis Disease in Fishes
I. PAPERNA and I. SABNAI (With 7 Figures) 228

Contributions to the Taxonomy of the Genus Diplozoon
von Nordmann, 1832
H.-H. REICHENBACH-KLINKE (With 21 Figures) 235

Treatment of Salmon Lice (*Lepeophtheirus salmonis* Kroyer)
with Neguvon
P.O. BRANDAL and E. EGIDIUS 248

Subject Index 249

Contributors

You will find the addresses at the beginning of the respective contribution

ACUIGRUP 206, 212
AHNE, W. 28, 186
ALDERMAN, D. 174
ATKINSON, H.M. 218
AUSTIN, B. 147
BAUDIN LAURENCIN, F. 60
BAUDOUY, A.M. 23
BEJERANAO, I. 224
BEKESI, L. 126, 127
BERG-BLOMMAERT, J.M. v.d. 120
BOOTSMA, R. 120
BRANDAL, P.O. 248
BUCKE, D. 82, 174
CHART, H. 39
CHILMONCZYK, S. 18
COURTICE, I.D. 218
CSABA, G. 126, 127
DANTON, M. 23
DELVENTHAL, H. 225
DIXON, P.F. 29
DORSON, M. 29
EGIDIUS, E. 248
ELLIS, A.E. 98
ENZMANN, P.-J. 185
EVELYN, T.P.T. 45
FAISAL, M. 186
GAYER, E.K. 127
HALLINGSTAD, F. 53
HARRISON, J.G. 193
HÅSTEIN, T. 53, 154
HASTINGS, T.S. 98
HILL, B.J. 29
JENSEN, N.J. 87
JØRGENSEN, P.E.V. 3, 8

KETCHESON, J.E. 45
KLINGER, H. 225
KÖRMENDY, B. 126
LARSEN, J.L. 87
LIVERSIDGE, J. 98
MCCARTHY, D.H. 113
MCVICAR, A.H. 165
MEIER, W. 8
MERLE, G. 23
MICHEL, C. 94
MOLNÁR, K. 157
MUNN, C.B. 39, 69
MUNRO, A.L.S. 98
NEUMANN, W. 90
PALMER, R. 107
PAPERNA, I. 228
PATERSON, W.D. 113
PETERS, G. 225
PLÖGER, W. 90
POL, J.M.A. 120
REFSTI, T. 53
REICHENBACH-KLINKE, H.-H. 235
RICHARDS, R.H. 75
ROALD, S.O. 53, 154
SABNAI, I. 228
SARIG, S. 224
SCHLOTFELDT, H.-J. 198
SCHULZ, D. 137
SEAGRAVE, C. 174
SMITH, P.D. 113
SMITH, P.R. 107
TANGTRONGPIROS, J. 60
TRUST, T.J. 218
WOOTTEN, R. 165

Viral Diseases

Chairman: P. DE KINKELIN, France

Egtved Virus: The Susceptibility of Brown Trout and Rainbow Trout to Eight Virus Isolates and the Significance of the Findings for the VHS Control

P. E. V. JØRGENSEN[1]

On the basis of limited experimental evidence (Rasmussen, 1965; Jørgensen, 1974; de Kinkelin et al., 1974) brown trout have generally been considered to be almost refractory to natural infection with Egtved virus. Nevertheless VHS in a few rare cases has been observed in brown trout under trout farm conditions. The author has isolated Egtved virus twice from such cases, once (1969) from brown trout of Italian origin, once (1972) from brown trout in a Danish trout farm (Jørgensen, unpublished results). Recently also de Kinkelin and Le Berre (1977) isolated Egtved virus from brown trout. The latter isolate, designated strain 23/75, was found to be pathogenic to brown trout as well as to rainbow trout in bath infection experiments.

The present experiments were carried out to make clear whether the pathogenicity of strain 23/75 to brown trout was a unique feature of that strain or whether other Egtved virus isolates were also brown-trout-pathogenic when tested by means of bath infection.

A total of eight virus isolates was examined, among others the reference strain of each of the three preliminarily identified serotypes, strain F1 (serotype 1), strain He (serotype 2; Jørgensen, 1972, 1974), and strain 23/75 (serotype 3; Le Berre et al., 1977).

Material and Methods

Groups of 20 brown trout (average body weight 6 g) and of 50 rainbow trout (average body weight 7 g) were exposed to virus concentrations of approximately 1×10^5 pfu per ml of water for one hour. Negative controls received cell culture medium instead of virus. The aquaria were supplied with running tap water at a temperature of $9°-10°C$. The observation period was six weeks, during which all dead fish were examined for the presence of VHS symptoms.

Results

The results of the experiments are shown in Table 1.

Among the eight virus isolates three were found to be pathogenic to brown trout as well as to rainbow trout. The remaining five were pathogenic only to rainbow trout.

---

[1] State Veterinary Serum Laboratory, Department for Jutland, Hangøvej 2, 8200 Århus N, Denmark

Table 1. Mortality in rainbow and brown trout after bath infection with 8 different strains of Egtved virus

Virus strain	Mortality in rainbow trout (%)	Mortality in brown trout (%)	Sero-type	50% pnt. against anti-F1[a]	Origin of virus strain	Virus inoculum produced in
F1	10	0	1	1,500	Rainbow trout	FHM cells
He[b]	64	0	2	500	Rainbow trout	–
23/75	64	58	3	300	Brown trout	–
Jedsted	68	0	?	1,200	Rainbow trout	–
Ansø	74	79	?	600	Brown trout	–
N.543	96	0	?	600	Rainbow trout	–
Bidstrup	62	0	?	600	Rainbow trout	–
No	58	32	?	300	Rainbow trout	–
Controls	0	0				

[a] 50% plaque neutralization titer
[b] Previously designated isolate 61 (Jørgensen, 1972)

Strain Ansø, which caused the highest mortality in brown trout, was, like strain 23/75, originally isolated from brown trout (Jørgensen, 1972, unpublished results), whereas strain No was isolated from rainbow trout.

The symptoms of the brown trout which died during the experiments were clearly those of VHS, and Egtved virus was readily isolated from affected individual fish. Serological identification of the reisolated virus strains was carried out by means of indirect immunofluorescence utilizing a rabbit antiserum to strain F1 (Jørgensen, 1974).

In 50% plaque neutralization tests all eight isolates were neutralized by a rabbit antiserum to strain F1 (Table 1).

Discussion

The results clearly show that brown trout are susceptible to some Egtved virus strains under the conditions of bath infection at a high virus concentration. Brown trout were not found to be susceptible to strain He as reported by de Kinkelin and Le Berre (1977). The reason for this discrepancy may be differences in the virus passage levels or virus cultivation techniques.

The above findings add an interesting detail to the knowledge of the epizootiology of VHS. In spite of the pronounced difference between the virus titer in the water during the bath infection experiments (10^5 pfu per ml) and during a natural infection (less than 1 $TCID_{50}$ per ml) (Jørgensen, 1974) it must probably be expected that brown trout can become VHS-infected in streams receiving the outlet from VHS-infected trout farms at least under conditions of severe stress. It thus appears that brown trout may make up a potential virus reservoir in the streams.

When the severity of the above findings with regard to VHS control in trout farms is estimated it has also to be considered that the wild fish population of many (perhaps

most) European rivers includes a large number of rainbow trout, which have either escaped from trout farms or been stocked into the rivers for fishery purposes.
Since rainbow trout are more susceptible to VHS than brown trout it must probably be expected that the quantitatively most important virus reservoir exists in the wild rainbow trout.

In addition to rainbow and brown trout, pike may also play a role as carriers of Egtved virus. At least this must be expected on the basis of a recent report by Meier and Jørgensen (1979) of a severe spontaneous outbreak of virologically verified VHS in pike fry in Switzerland. The report in question represents the first demonstration of VHS in nonsalmonid fish, and it may justify the feeling that more species of fish may in time turn out to be susceptible to the disease.

In rivers with several VHS-infected trout farms situated next to each other along the stream and with free movement of wild fish up and down the stream it is likely, considering the susceptibility of several species of fish to the virus, that a smaller or larger percentage of the wild fish are, at least periodically, harboring the virus. Against that background it might appear necessary to remove the wild fish population before prevention of VHS in the trout farms is attempted, since persistence of the infection in the wild fish might lead to reinfection of sanitized trout farms.

In Denmark control of VHS in trout farms has been practiced for more than 15 years without measures being taken to remove the wild fish populations (Jørgensen, 1974). The basis for this policy was the recognition that it would be very difficult or impossible to obtain permission for such steps from government authorities and private owners and that it would be extremely difficult to remove all wild fish from a stream, even if permission was obtained.

The principle of the VHS control program is that all VHS-infected farms along a stream, starting at the top of the stream, are emptied, disinfected, kept dry for one

		A	B	C	D	E	F	G
A.								
	1968	0	+	+	+	+	+	+
B.	1969	0	E+	E+	+	+	+	+
	1970	0	E	E	+	+	+	+
C.	1971	0	0	0	E	E	E	E
	1972	0	0	0	0	0	0	0
	1973	0	0	0	0	0	0	0
D.	1974	0	0	0	0	0	0	0
	1975	0	0	0	0	0	0	0
E.	1976	0	0	0	0	0	0	0
	1977	0	0	0	0	0	0	0
F.	1978	0	0	0	0	0	0	0
	1979	0	0	0	0	0	0	0
G.								

+ VHS infection; *E* "stamping out" procedure; 0 indicates freedom from VHS; *Two signs* indicate two events in one year

Fig. 1. Development of the VHS control program in 7 trout farms (A–G) at the river "Lille Å" (The figure is based on information obtained from Dr. N. P. Kehlet, Veterinary Services, Copenhagen)

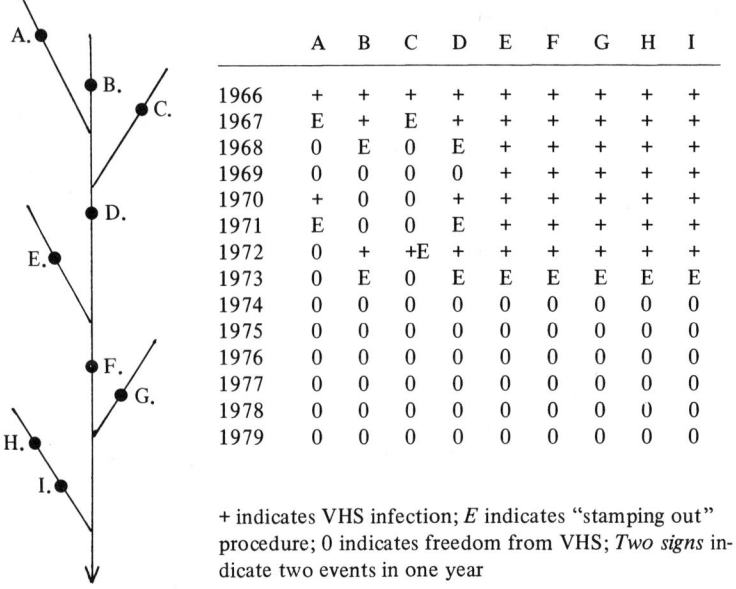

	A	B	C	D	E	F	G	H	I
1966	+	+	+	+	+	+	+	+	+
1967	E	+	E	+	+	+	+	+	+
1968	0	E	0	E	+	+	+	+	+
1969	0	0	0	0	+	+	+	+	+
1970	+	0	0	+	+	+	+	+	+
1971	E	0	0	E	+	+	+	+	+
1972	0	+	+E	+	+	+	+	+	+
1973	0	E	0	E	E	E	E	E	E
1974	0	0	0	0	0	0	0	0	0
1975	0	0	0	0	0	0	0	0	0
1976	0	0	0	0	0	0	0	0	0
1977	0	0	0	0	0	0	0	0	0
1978	0	0	0	0	0	0	0	0	0
1979	0	0	0	0	0	0	0	0	0

+ indicates VHS infection; E indicates "stamping out" procedure; 0 indicates freedom from VHS; *Two signs* indicate two events in one year

Fig. 2. Development of the VHS control program in 9 trout farms (A–I) at the river "Simested Å" (The figure is based on information obtained from Dr. N. P. Kehlet, Veterinary Services, Copenhagen)

month, and then repopulated with fish from VHS-free farms. As shown in Figs. 1 and 2 there are cases in which this policy has been very successful, since it has led so far to freedom from VHS in all the farms involved for periods of 8 and 6 years respectively.

It is not known whether in those cases the wild fish in the streams were actually carriers of Egtved virus or not. If they were it might be suspected that the duration of the virus carrier state was short, since the VHS eradication program was successful.

Probably the removal of the common sources of virus, i.e., the densely populated VHS-infected trout farms, caused a drop in the virus concentration in the streams as a side effect of the sanitation. Perhaps the virus concentration reached a level at which new infection of wild fish did not occur and at which already infected fish were able to clear themselves of the infection.

Since there is no experimental evidence available which supports the above hypothesis it is put forward only as a working theory. In other streams the VHS control program has been less successful, since reinfection has sometimes occurred shortly after the repopulation of the farms. It is not clear at present to what extent this has been due to persistence of the infection in the wild fish and to what extent to other sources of virus such as closely situated VHS-infected farms from which virus may be transferred by animals, personnel, trucks, and so on.

It would be very valuable to have experiments initiated which could help in estimating the relative significance of each of those two fundamentally different causes of VHS reinfection in sanitized trout farms.

References

Berre M Le, Kinkelin P de, Metzger A (1977) Identification sérologique des Rhabdovirus des salmonidés. Proc 2nd Copraq Meet (FAO/EIFAC – OIE), Brest 1977, pp 390–393
Jørgensen PEV (1972) Egtved virus: Antigenic variation in 76 virus isolates examined in neutralization tests and by means of the fluorescent antibody technique. In: Mawdsley-Thomas (ed) Symp Zool Soc London No 30, Diseases of fish. Academic Press, London New York, pp 333–340
Jørgensen PEV (1974) A study of viral diseases in Danish rainbow trout, their diagnosis and control. Thesis, pp 101, commissioned by A/S C.F. Mortensen, Bülowsvej 5c, 1870 Copenhagen V
Kinkelin P de, Berre M Le (1977) Isolement d'un Rhabdovirus pathogène de la truite Fario (Salmo trutta). C R Acad Sci Ser D 284:101–104
Kinkelin P de, Berre M Le, Meurillon A (1974) Septicemie hemorrhagique virale: Demonstration de l'etat refractaire du saumon (Oncorhynchus kisutch) et de la truite Fario (Salmo trutta). Bull Fr Piscic 253:166–176
Meier W, Jørgensen PEV (1979) Isolation of Egtved virus from pike fry (Esoc lucius) with hemorrhagic symptoms. Proc 3rd Copraq Meet (FAO/EIFAC – OIE), Munich 1979 in press
Rasmussen CJ (1965) A biological study of the Egtved disease (INUL). Ann N Y Acad Sci 126: 427–460

Isolation of VHS Virus from Pike Fry (Esox lucius) with Hemorrhagic Symptoms

W. MEIER[1] and P.E.V. JØRGENSEN[1]

In the early summer of 1978 we received pike fry for examination because of a disease outbreak in a hatchery on the Hallwilersee, an average-sized inland lake, in Switzerland. The fry, which originated from Hallwilersee, had been kept on a plankton diet for 3 weeks in the hatchery, in filtered lakewater, when the first signs of disease occurred. During a seven-day period, the hatchery lost a total of about 120,000 pike fry; close to 100% of the population. The mortality occurred shortly after an increase of the water temperature from 12 °C to 14 °–15 °C. In most circumstances the time between onset of symptoms and death was less than 2 days. During this hatching period no other species of fish were present in the hatchery. The fry, 2.5–3.5 cm long, had overall exophthalmus and pronounced reddening of the flanks and, less frequently, the middle of the skull. The gills were very pale. Dissection of the fry revealed ascites and hemorrhages in the kidney area.

The whole fish or the kidney, liver, and spleen of the freshly killed fry, with typical symptoms of the disease, were virologically examined by inoculation into RTG-2 and FHM cells, following homogenization, dilution, and filtration through a 450-mμ filter. Inoculated cell cultures were incubated at 15 °C.

After 3–4 days, a clear CPE was observed in both cell types. A routine serological identification test (neutralization test in tube cell cultures) revealed that the cytopathic agent was neutralized by antiserum to Egtved virus, strain F1.

To confirm this result, the new pike virus was examined in plaque neutralization tests and by indirect immunofluorescence (Jørgensen, 1974) against the above-mentioned antiserum to Egtved virus. For comparison, the following fish rhabdoviruses were included in the tests: three isolates of Egtved virus, representing three proposed serotypes of the virus (F1, He, and 23/75), the pike fry rhabdovirus (PFR), rhabdovirus carpio (RVC), and infectious hematopoietic necrosis virus (IHN). As shown in Table 1, the new pike virus was clearly related to Egtved virus and nonrelated to the PFR, RVC, and IHN viruses.

The pathogenicity of the new pike virus to pike and rainbow trout was compared, respectively, to that of a freshly isolated Egtved virus strain originating from rainbow trout. Artifical infection was carried out as bath infection and partially by intraperitoneal injection, as shown in Fig. 1.

The virus used in the infection experiments was produced in FHM cells, utilizing virus of the second cell culture passage as inoculum. Each batch of virus was titrated in

---

1 Untersuchungsstelle für Fischkrankheiten, Vet. bakt. Institut der Universität Bern, Länggasstr. 122, CH-3012 Bern
2 State Veterinary Serum Laboratory, Department for Jutland, Hangøvej 2, 8200 Århus N, Denmark

Table 1. Pike virus and other Rhabdovirus in reaction with rabbit anti F-1

	Reaction with Rabbit Anti F-1	
	50% Plaque Neutr. Titre	IFAT
Pike-Virus	800	+
F-1	1500	+
He	500	+
23/75	300	+
PFR	< 20	o
RVC	< 20	o
IHN	< 20	o

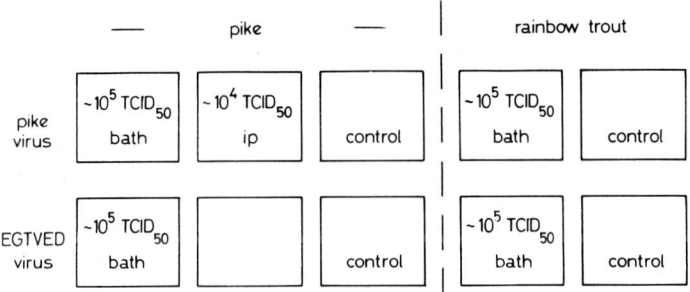

Fig. 1. Infection protocol

microtiter trays and frozen at −20 °C until use. Virus titers were calculated according to the Kärber method.

Each aquarium used in the experiment had a volume of 200 l and was supplied with tap water. The temperature variation was 12 °C–15 °C. Each aquarium contained 100 fish and strict hygiene was maintained during the experiment.

Pike measuring 2–3 cm were obtained from Moossee. They had been fed plankton for two weeks up to the time of transfer to the laboratory. Before the experiment they were acclimatized for 7 days and throughout the experiment they received fresh plankton from Moossee daily.

The rainbow trout were obtained from a spring-water-supplied hatchery with a known history of VHS freedom, and measured 4.5–5.5 cm. They received pelleted trout food throughout the experiment.

In both species, each of the two viruses caused acute mortality between the 3rd and 14th day. On the latter day 80%–90% of all groups of infected fish had died. The noninfected control trout had no mortality, however a 6% mortality due to weakness, starvation, and injury was observed among the noninfected control pike. The results of the infection experiments are shown in Figs. 2–4.

Virus was readily isolated from fish belonging to the artificially infected groups, but not from noninfected control fish.

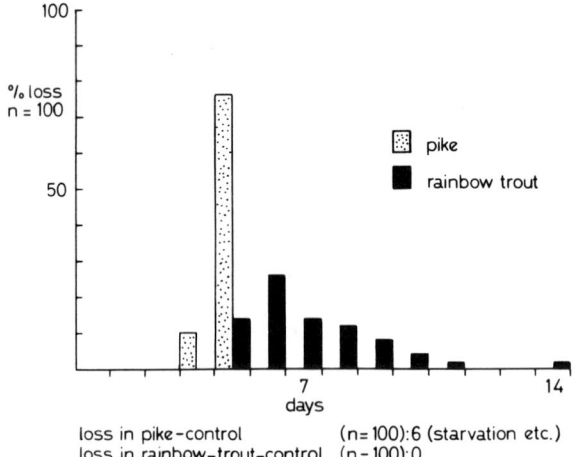

Fig. 2. Loss after infection with Pike virus

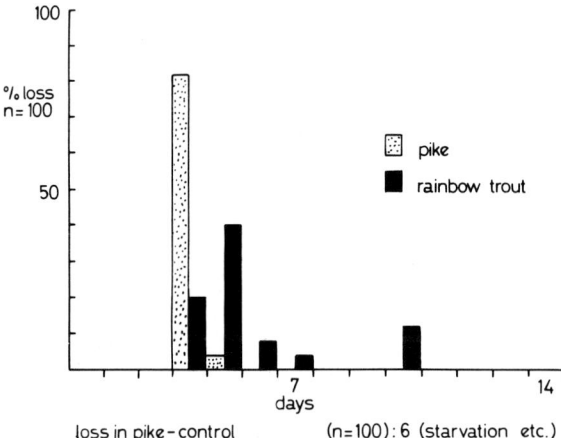

Fig. 3. Loss after infection with Egtved virus

The symptoms of the diseased rainbow trout (regardless of the virus type used for the infection) were evidently those of acute viral hemorrhagic septicemia (VHS).

The two viruses caused identical symptoms in the pike. Since VHS has not been previously described in pike, it will be briefly described here. The fish became apathic and floated on their side at the water surface or remained at the bottom of the aquaria. Typically swelling and reddening could be observed in the flanks and in the base of the breast- and tail fins; less frequently hemorrhages were evident on the top of the head; a slight bilateral exophthalmus was often observed; the gills were extremely pale; the abdomen mostly was filled with a clear, slightly yellowish or sometimes even hemorrhagic fluid; the intestinal tract usually was full of a milky slime; the liver was of a white-greyish colour.

A detailed study of histological and electron microscopical changes in the tissue of fish infected with each of the two viruses will be published elsewhere. Very briefly it

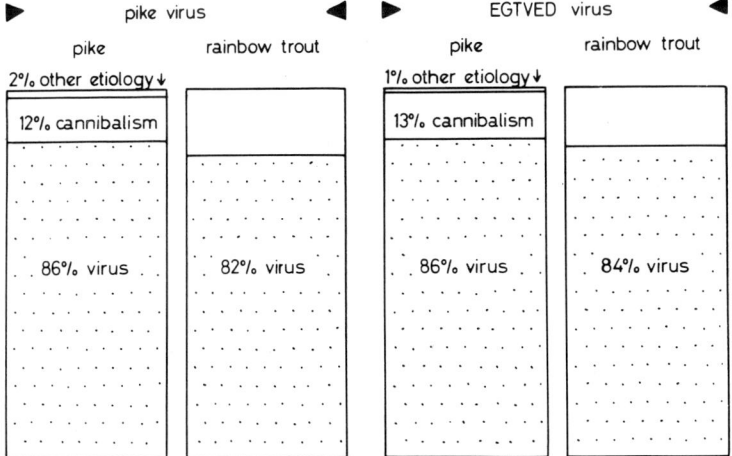

Fig. 4. Total loss over 14 days

shall be mentioned that the lesions induced by each of the two viruses, apart from a slight difference in intensity, were very similar. The changes induced by each of the viruses were identical whether in pike or in rainbow trout, although the hemorrhages tended to be more violent in the pike.

Massive hemorrhages were present subperitoneally in the kidney region and intermuscularly in the tail area, particularly in pike. Hemorrhages of a lower degree were observed in other parts of the musculature. They were mostly accompanied by myolysis.

Other characteristic findings were hemorrhages and edema, often with detaching of the retina within the retro-bulbar space. Epidural hemorrhages were often present in the tectum opticum area, less frequently in other parts of the nervous system.

Very severe degenerative changes in the tubular epithelial cells, sometimes even necrosis, were the most dominant feature in the kidney. Different stages of nuclear degeneration were observed. In the cytoplasm irregularly sized granules reminded one of inclusion bodies. The tubular lumen often filled with necrotic material and hyaline casts, and the glomerula showed a slightly edematous mesangium. In the interstitium there were focal hemorrhages, edema, and necrosis of individual cells.

The ultrastructural examination of renal tubular epithelial cells confirmed the degenerative lesions described. The suspected inclusion bodies turned out to be cytoplasmic debris besides a lot of partly swollen mitochondria. More altered cells often showed conglomerates of altered mitochondria in lysosomes.

The slight changes of the liver were characterized by a congestion with edema and small necrotic foci or single-cell necrosis. There were also small necrotic foci in the pancreas and the spleen. The other organs showed no specific changes.

The results show that pike are susceptible to VHS and that the disease can cause extremely severe losses in fish other than salmonids. In pike fry the losses were close to 100% in the experiments as well as during the natural outbreak of the disease. VHS in pike fry, thus, is a disease of practical significance. It is still unknown whether pike

Fig. 5. a Wide-spread subperitoneal hemorrhages in the distal kidney area. HE; medium magnification (mM)/Pike virus (PV) in pike (P). **b** Focal intermuscular hemorrhages. HE; mM/PV in P

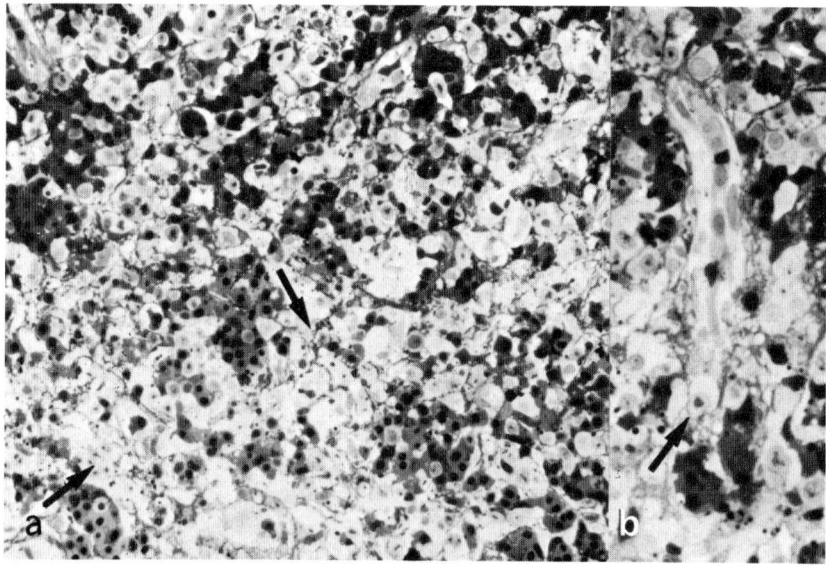

Fig. 6. a Disseminated necrosis (↑) in the renal interstitium. Tol. blue; mM/PV in P. **b** Higher magnification of **a**. Degenerative cells of the sinus show pyknotic nuclei (↑). Tol. blue; high magnification (hM)/PV in P

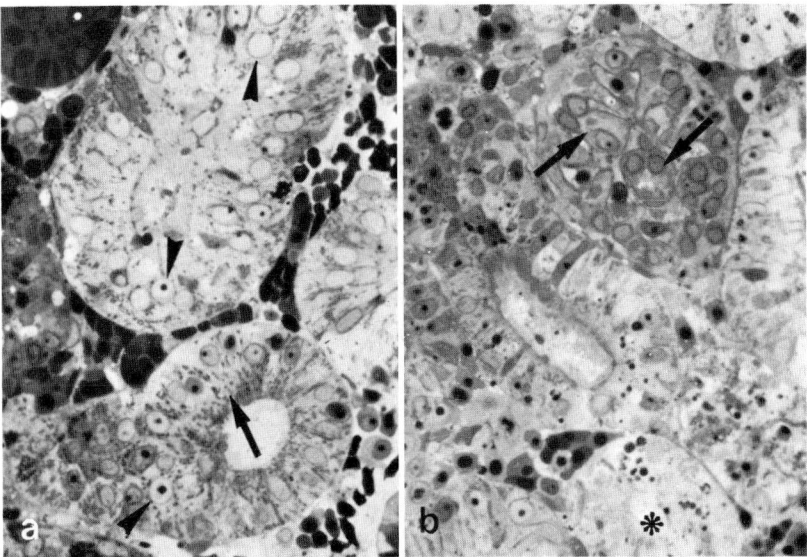

Fig. 7. a Renal tubular epithelial cells show typical cytoplasmic granules (↑) and degenerative nuclear changes (▲). Tol. blue; hM/PV in P. **b** Apart from necrotic tubules (∗) the glomerula show a swollen mesangium (↑). Tol. blue; hM/PV in P

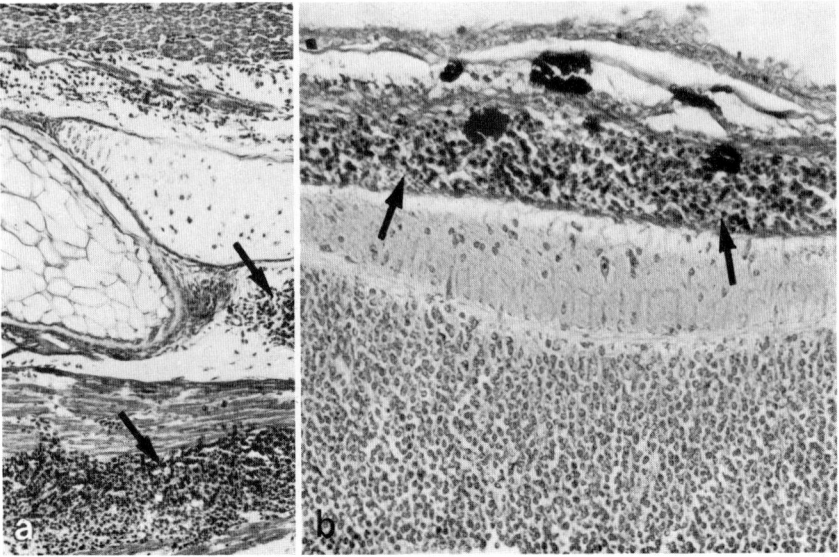

Fig. 8. a Hemorrhages (↑) in the tail area. HE; low magnification (lM)/Egtved virus (E) in P. **b** Epidural hemorrhages (↑) in the tectum opticum region. HE; lM/E in P

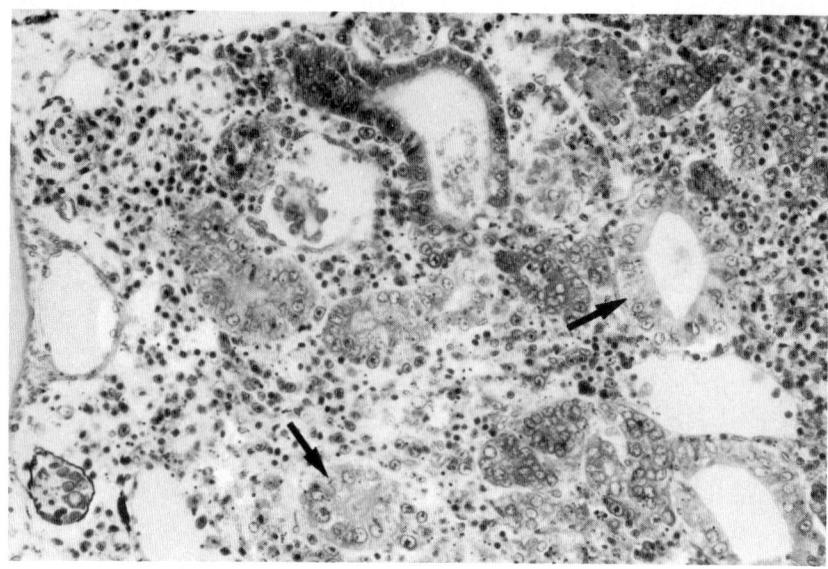

Fig. 9. Disseminated necrosis in the renal interstitium and various degrees of degenerative tubular lesions (↑). HE; mM/E in P

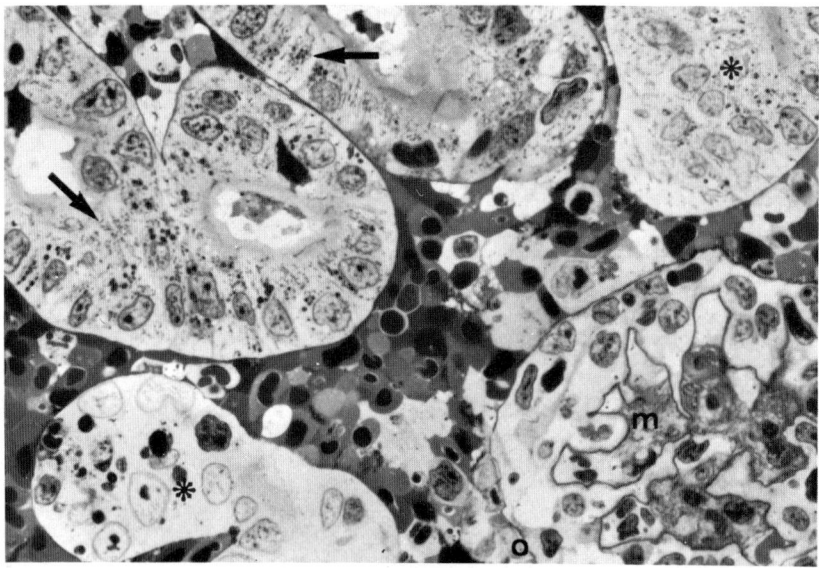

Fig. 10. Slight swelling of glomerular mesangium (*m*); tubular cells show typical cytoplasmic granules (↑) and are partly necrotic (∗). Nuclei become hyperchromatic and pyknotic, hyaline casts are visible. Tol. blue; hM/PV in rainbow trout (Rb)

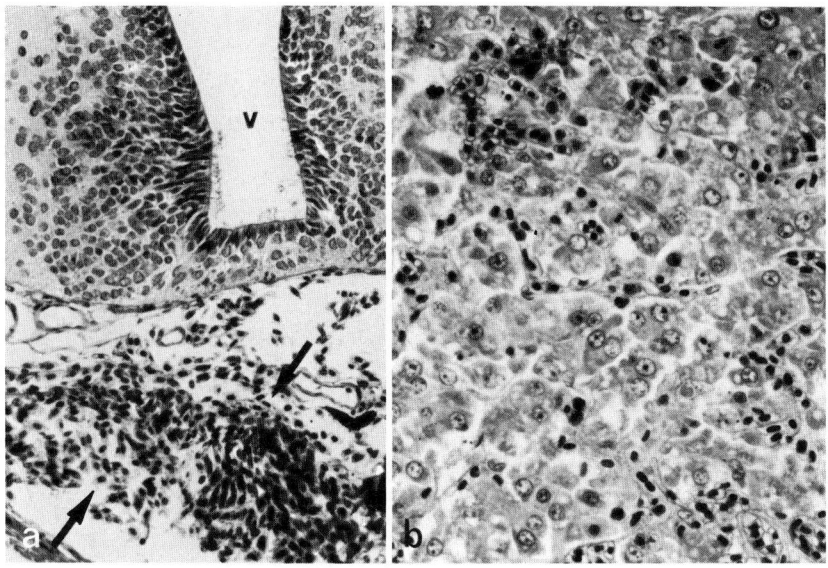

Fig. 11. a Hemorrhage (↑) in the tectum opticum region. *v*, 3 rd ventricle; HE: lM/PV in Rb. **b** Focal liver necrosis. HE; mM/PV in Rb

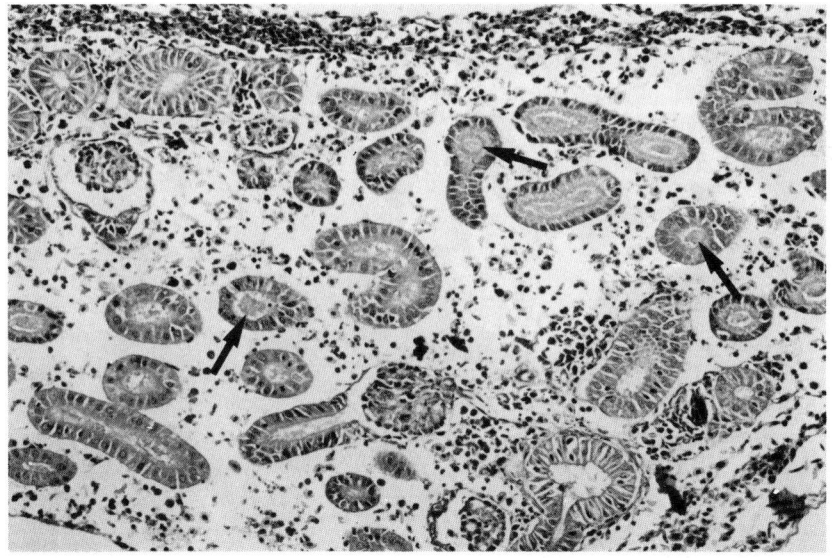

Fig. 12. Nephrotic area, intratubular casts are visible (↑). HE; lM/E in Rb

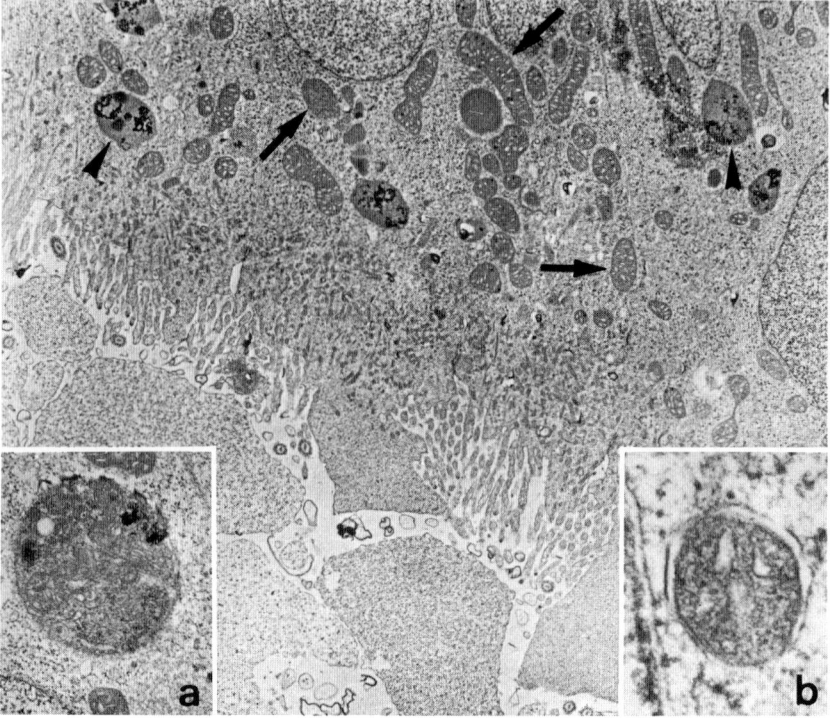

Fig. 13. Prominant, partly swollen mitochondria (↑) of various size and many residual bodies (▲) within degenerative renal tubular cells. x6,880. The *insets* show a higher magnification of: *(a)* a conglomerate of phagocytized, altered mitochondria within a lysosome (x12,060), and *(b)* a swollen mitochondrion (x54,120)

at later stages of development can be attacked by the disease or become carriers of the virus. It is evident, however, that as far as the epizootiology of VHS is concerned, pike may play an important, previously completely unexpected role. This is true particularly in countries like Switzerland where pike are distributed artificially, not only for sport fishery, but also with the aim of maintaining an ecological population balance. In Switzerland, several million pike fry and several tons of older pike are stocked every year in a large number of appropriate lakes. Pike may thus contribute to a dissemination of virus in previously noninfected areas, and in hatcheries.

An important problem in the future will be how to distinguish PFR disease (PRFD) from VHS of pike fry. PFRD, according to Bootsma (1976), attacks pike at the same age and produces the same symptoms and apparently almost identical pathoanatomical and histopathological changes.

The basis for the differentiation between the two diseases apparently has to be virus cultivation plus some sort of virus identification.

Whereas serological identification of PFRD virus remains a very difficult task (Bootsma, 1978, personal communication), serological identification of Egtved virus is routinely performed in several laboratories. In the present work it was discovered that one single antiserum (against the reference strain F1) was sufficient for a positive identification of all examined Egtved virus isolates including the one from pike. The same antiserum allowed Egtved virus to be clearly differentiated from other fish rhabdoviruses. This was true whether the technique applied was plaque neutralization or indirect immunofluorescence, and since the latter technique is simple and rapid it seems to be the choice method for primary Egtved virus identification. Ideally, this would be followed by a plaque neutralization test against serotype-specific anti-Egtved sera.

References

Bootsma R (1976) Studies on two infections diseases of cultured freshwater fish. Rhabdovirus Disease of Pike Fry (Esox lucius L). Columnaris Disease of carp (Cyprinus caprio L). Thesis University of Utrecht, p 41

Jørgensen PEV (1974) Indirect fluorescent antibody techniques for demonstration of trout viruses and corresponding antibody. Acta Vet Scand 15:198–205

Some Aspects of Trout Gill Structure in Relation to Egtved Virus Infection and Defence Mechanisms

S. CHILMONCZYK[1]

Introduction

The natural pathway of viral hemorrhagic septicemia (VHS) virus penetration in trout is unknown. Fish pathologists suspect the gill as a possible route for virus entry into the fish. Some results obtained with another rhabdoviral model, the spring viremia of carp, showed that the gill can play an important role in viral pathogeny [1].

In trout, the entry probably takes place at the secondary lamella level. Gill arch and gill filament epithelia are several layers thick and contain a large number of protective mucous cells, but in secondary lamella we find only a bilayered epithelium with a small number of mucous cells (Fig. 1). In secondary lamella the blood flows between pillar cells (PC) which delimit the blood space completely. So blood is isolated from water only by a very thin wall composed of two layers of epithelial cells and cytoplasmic flanges of PC (Fig. 1).

In the present work, using virological and cytological methods we have studied the trout fill during and after VHS experimental infections.

Material and Methods

Fish

Rainbow trout weighing 1 g or 180 g were maintained in tanks supplied with dechlorinated tap water at a temperature of 11 ° ±1 °C.

Viral Infections

180 g fish were infected by a 0.2-ml intracardiac injection of VHS type 1 (Egtved) virus (10^8 pfu/fish) mixed with latex spherules (0.81 μm; DIFCO, Detroit, USA); 1 g fish were maintained 3 h in the infectious bath (10^8 or 1.5×10^9 pfu/ml water) for a waterborne contamination.

Sampling: after infection, gill, kidney, and spleen were sampled for cytological and virological examinations at various intervals from 5 min to 72 h.

---

1 Laboratoire d'Ichtyopathologie, Institut National de la Recherche Agronomique, 78850 Thiverval-Grignon, France

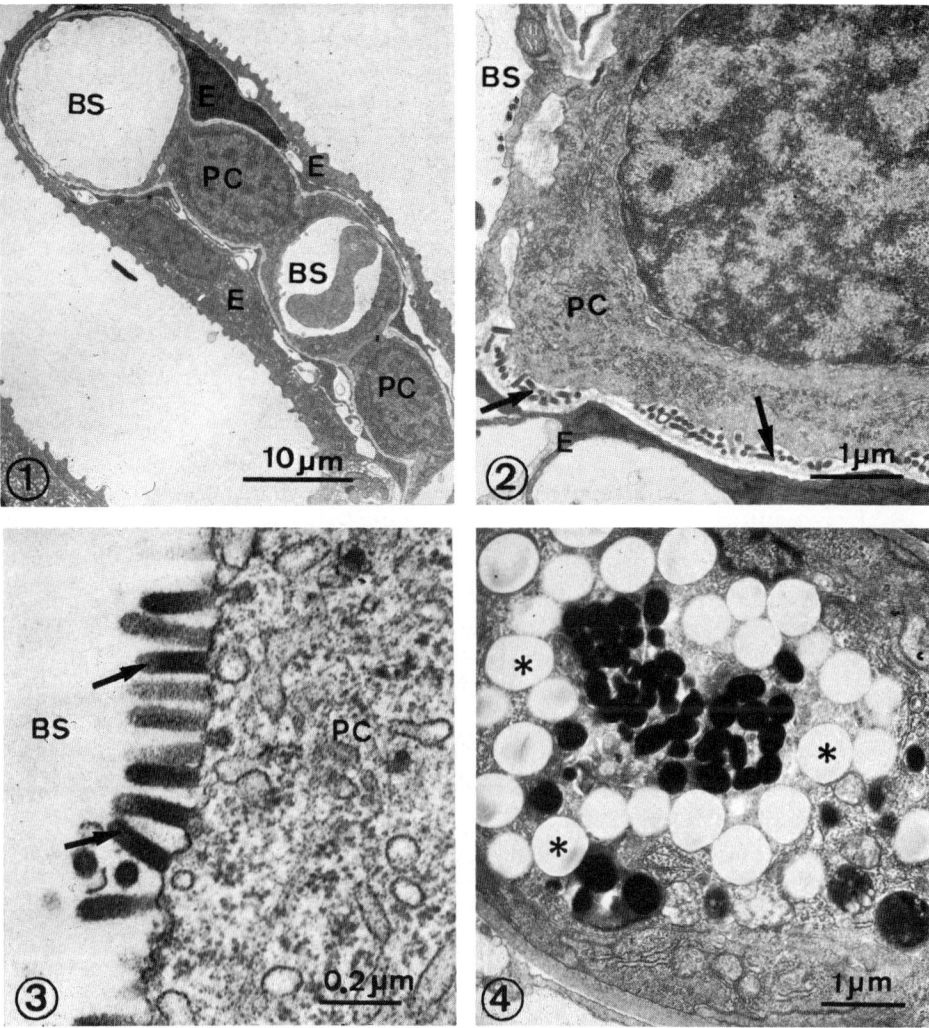

Fig. 1. Electron micrograph of one secondary lamella. Blood spaces are delimited by pillar cell flanges. *BS* blood space; *E* epithelial cell; *PC* pillar cell

Fig. 2. Electron micrograph showing a section of secondary lamella (72 h post infection) with an infected pillar cell surrounded by budding VHS virus particles *(arrows)*. Abbreviations as in Fig. 1

Fig. 3. Electron micrograph of budding VHS particles *(arrows)* 72 h post infection. Viruses are directly scattered into blood space. Abbreviations as in Fig. 1

Fig. 4. Electron micrograph of pillar cell 24 h after VHS virus-Latex mixed injection. Numerous latex spherules (∗) are ingested. The cells contains dark intracytoplasmic pigment

Cytological Examinations

tissues were fixed at 4 °C for 2 h in 1.6% glutaraldehyde buffered with sodium cacodylate (ph 7.3), then post-fixed for 1 h in osmium tetroxide and embedded in Epon. Ultrathin sections were stained with uranyl acetate and lead citrate according to the Reynolds method [8]. Semithin sections (1 μm) for light microscope observation were stained with methyl-blue and Azur II.

Virological Examinations

VHS cytopathic effect was revealed according to routine techniques [3]. RTG_2 monolayers in 24 wells Costar plates were inoculated with 0.1 ml of organs extracts diluted from 1/20 to $1/2x10^5$.

Results

Waterborne Infections

Similar results were obtained with fry infected either by 10^8 or by $1.5x10^9$ pfu/ml of contaminated water (Fig. 5). Five minutes after the beginning of infection, virus was detected only in gill by virological techniques, but the direct penetration into this organ was not demonstrated with the electron microscope. Afterwards virus could not be detected in the gill for 3 days, after which virus was again detectable in gill by virological techniques (dilution 1/200). On day 3 post infection the electron microscope showed that the only infected gill cells were the PC (Fig. 2). At that time the disease was in its septicemic period and virus, scattered into the blood stream (Fig. 3), was found also in spleen and kidney.

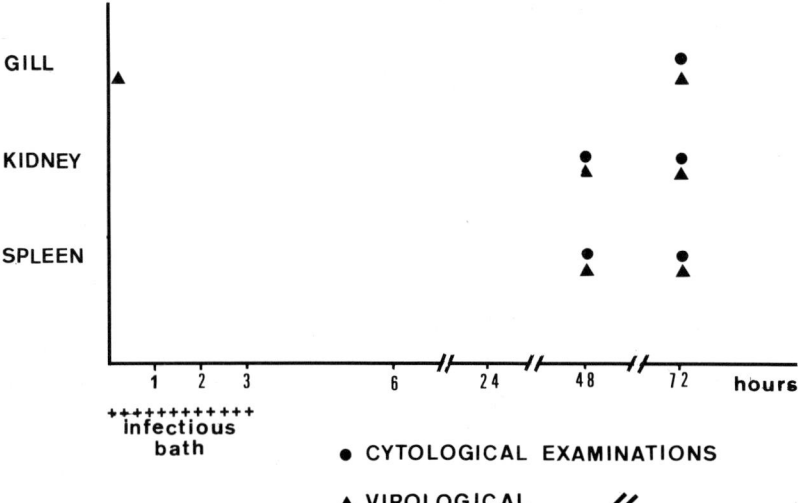

Fig. 5. VHS waterborne infection: 10^8 or $1.5x10^9$ pfu/ml

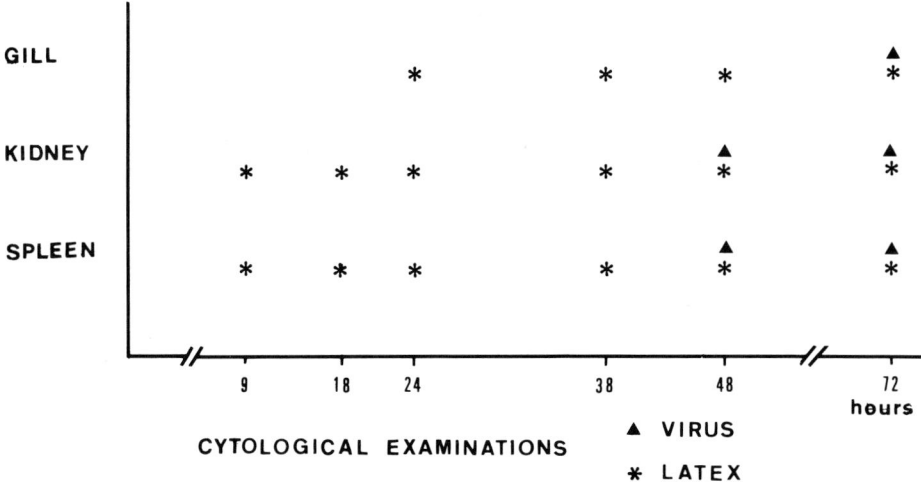

Fig. 6. VHS virus-latex mixture intra-cardiac injections (10^8 pfu/fish)

VHS Virus – Latex Mixture Injections

Samples were examined only in the electron microscope (Fig. 6). An important phagocytic activity takes places rapidly in kidney and spleen. The uptake of foreign material mainly occurs in these two organs. In gill only the PC have a phagocytic activity (Fig. 4). Latex spherules ingestion requires a longer time in PC (24 h) than in kidney or spleen (9 h). In some phagocytic PC we observed intracytoplasmic pigment (Fig. 4). As in the case of waterborne infection, PC were the only gill cells in which virus replication could be observed. We have noticed that infected PC never contained latex spherules or pigment inclusion. In the same way latex-phagocytized PC displayed no evidence of viral replication.

Discussion

Fish gill lamella was studied mainly at the epithelium level for its respiratory [4] and osmoregulatory function [6]. Pillar cells which can synthesize collagen are considered supporting epithelial cells [2]. In this study we show that after latex injection into the blood stream some PC are involved in blood clearance. The melanotic PC ressemble the melanotic macrophages [7] generally found in spleen and kidney. The phagocytic PC represent a small part of the PC population. We can estimate that less than 4% of the number have phagocytic properties. Our observations are still too limited to determine the genuine role played by the cells in fish defence.

Carp gill was suspected as a possible target for spring viremia virus penetration [1]. Here, using cytological techniques, we have determined the exact site of viral replication into the gill: the PC; but virus penetration through the gill epithelium could not be demonstrated. This can be due to the fact that the number of virus particles which infect a fish is probably limited. Recently it was shown that 5–10 pfu are sufficient to kill a young fish [5], so visualization remains hazardous.

It must be stressed that we were unable to find virus-replicating PC displaying phagocytic properties at the same time. No obvious cytological differences allow us to distinguish these cells. It seems that presence or absence of pigment is not a clue for phagocytic or replicating activity. Indeed we have noticed that some phagocytic PC display no pigment. We can only suppose that a virus-infected cell loses its phagocytic properties. An alternative explanation is suggested by preliminary in vitro experiments. Blood polynuclear cells and kidney macrophages were cultured with virus latex mixture. Latex was ingested by the cells but the virus did not grow in them, supporting the view that phagocytic cells are not target cells for VHS virus. Although the relation between phagocytosis and viral replication in PC is not clear, this study reveals these two important aspects of the PC role in pathogeny. So these results reinforce the hypothesis that the gill plays a critical role in the first steps of disease processes.

References

1. Ahne W (1978) Uptake and multiplication of Spring Viraemia of Carp Virus in Carp *Cyprinus carpio* L. J Fish Dis 1:265–268
2. Bettex-Galland M, Hugues GM (1972) Demonstration of a contractile actomyosinlike protein in the pillar cells of fish gills. Experientia 28:744
3. Ghittino P, Kinkelin P de (1975) Méthodes d'interventions proposées contre les principales maladies contagieuses des poissons. Bull Off Int Epiz 83:649–687
4. Hugues GM, Morgan M (1973) The structure of fish gills in relation to their respiratory function. Biol Rev 48:419–475
5. Kinkelin P de, Chilmonczyk S, Dorson M, Le Berre M, Baudouy AM (1979) Some pathogenic facets of Rhabdoviral infection of salmonid fish. In: Bachmann PA (ed) Munich Symposia on Microbiology: Mechanisms of viral pathogenesis and virulence, p 357
6. Maetz J (1971) Fish gills: mechanisms of salt transfer in fresh water and sea water. Philos Trans R Soc Lond B 262:209–249
7. Roberts RJ (1975) Melanin containing cells of Teleost fish and their relation to disease. In: Ribelin WR, Migaki G (eds) Anatomic pathology of fishes. University of Wisconsin Press, Madison, p 399
8. Reynolds ES (1963) The use of lead citrate at high pH as an electron-opaque stain in electron microscopy. J Biophys Biochem 17:208–212

Experimental Infection of Susceptible Carp Fingerlings with Spring Viremia of Carp Virus, Under Wintering Environmental Conditions

A.M. BAUDOUY, M. DANTON and G. MERLE[1]

The rhabdoviral infection of carp has been called spring viremia (SVC) [1] because of its spring burst. So the aim of the present work was to investigate the infection processes occurring during winter in order to bring some light to the triggering of the disease.

Materials and Methods

Fish

The fish were SVC-virus-free carp, weighing around 50 g each, stored in 40-l-aquariums supplied by dechlorinated tap water flowing through.

Virus

The SVC virus strain was supplied by Dr. Pfeil-Putzien. It was produced in Epithelioma Papulosum Cyprini (EPC) cell line, provided by Professor Fijan, grown in Stoker's medium.

The conditions for the EPC cell cultivation, the virus titration by a plaque assay technique and the seroneutralization test were the same as those previously described using Fathead Minnow cells [2].

For screening the samples for their virus content, the cells were set in culture plastic plaques (24 wells/plaque, COSTAR) and each well was inoculated with a volume of 0.1 ml before receiving 1 ml of Stoker's medium and being incubated at +20 °C for 1 week. The cultures were observed daily.

Experimental Design

Each experimental group involved 75 tagged fish (25 fish per aquarium). For the water route infection trials, 2 aquariums No. 1 and No. 2 were inoculated at the final concentration of 5×10^3 pfu per ml water for 2 h, and uninfected cell culture fluid was added to the third aquarium used as a control.

---

[1] Ministère de l'Agriculture – Direction de la Qualité – Services Vétérinaires
Laboratoire Central de Recherches Vétérinaires – 22, rue Pierre-Curie, B.P. n° 67, 94703 Maisons Alfort Cédex, France (Directeur: L. Dhennin)

In the intraperitoneal (IP) experiments, the carp from 2 aquariums (No. 1 and No. 2) received 5×10^3 pfu each in 1 ml of medium and the controls received an equal volume of uninfected cell culture fluid.

Then, the subsequent mortalities were monitored over a seven-month period. The fish were also checked during this time for gill infection, viremia, virus shedding, and neutralizing antibodies.

The samplings were performed alternately on 20 fish of the aquarium No. 1 and No. 2 of each group on the following days post-infection:
Aquariums No. 1: 1, 3, 7, 14, 30, 75, 105, 135, 165
Aquariums No. 2: 2, 4, 10, 21, 60, 90, 120, 150, 180.

In each infected group, 5 fish were left unbled to ensure that the mortalities were not due to the bleeding.

Gill Biopsy

The ground end of a 1-ml pipette was passed over the surface of gill filaments until a small amount of material could be aspirated. This material includes a liquid and a small number of cells. Then, the sample is diluted 1/40 in Stoker's medium and inoculated to EPC cells. So the same fish could be checked up to 9 times without exhibiting evident troubles.

Virus Shedding

Virus shedding detection was achieved by storing the infected fish for 1 h individually in a plastic bag containing 100 ml sterile water. Afterwards the virus content of water was revealed by inoculation to cells.

Results and Discussion

Figures 1 and 2 show that the variations of the temperature values, from +11 °C to +5 °C, then from +5 °C to +20 °C, simulate well those of wintering conditions, followed by the spring temperature rise.

Water Route Contamination (Fig. 1)

The gill infection curve indicates that the water route infection procedure enables the contamination of all the fish to take place. Indeed, 50% of them are virus-positive on day 2 post infection and all of them on day 3.

On the other hand, the gill biopsy constitutes a non traumatic reliable technique to screen a fish stock for SVC. It shows that the percentage of gill-infected carp always remains higher than 80% until day 120. Then, it starts dropping from 100% to 0 within 1 month and most of the fish have died by them.

Compared to the results reported by Ahne [3] two years ago, the virus detection in gill occurs earlier: on the 2nd day instead of the 4th and so does the viremia, which

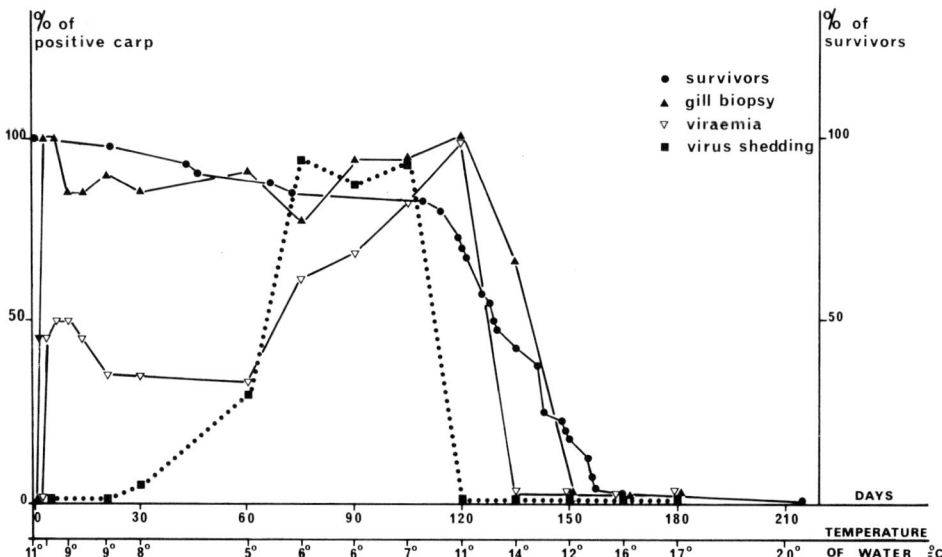

Fig. 1. Results of waterborne infection of 50-g carp following a 2-h exposure to 5×10^3 pfu SVC virus per ml water at +11 °C. Water temperature decreased to +5 °C within 60 days after infection and then rose gradually to reach +20 °C on day 210. Each curve represents the individual responses of two groups of 20 fish, the samplings for one group being made every other time on indicated days

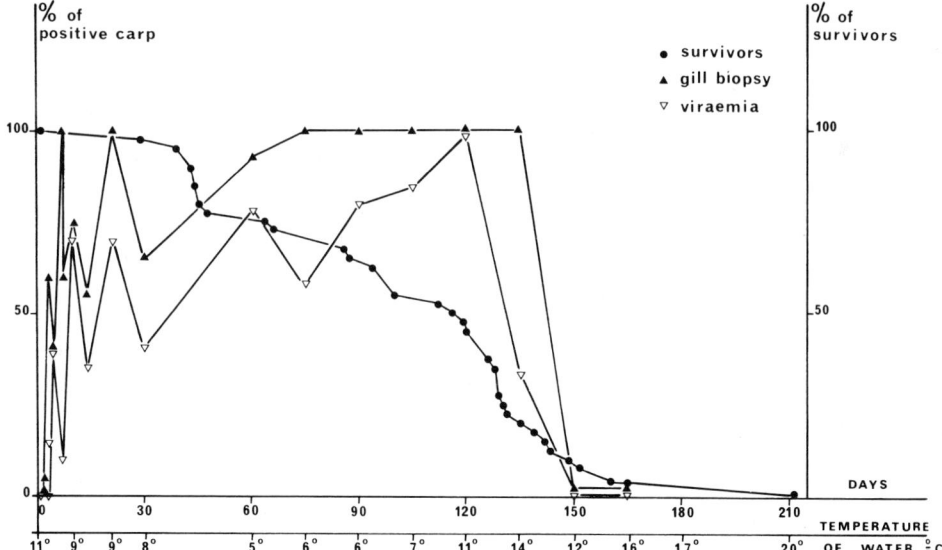

Fig. 2. Results of infections trials of carp injected intraperitoneally with 5×10^3 pfu SVC virus per animal. Same thermal variations and samplings days as Fig. 1

appears here on day 3 instead of day 6. These differences can be explained by the fact that we have been working with 40 individual samples instead of pools. Anyway, these results substantiate the possibility that gill could be the first site of the virus multiplication.

Despite the rapid contamination of all animals, the survival curve goes down slowly during the decreasing temperature phase. But all the fish which died during this time exhibited the typical clinical signs of SVC and for all of them the respective viremia values had been increasing until their death. This demonstrates that clinical SVC can occur at very low temperatures.

When the gill infection was present in most of the fish, the viremia remained detectable in less than 50% of them, until day 60 it was revealed by a number of plaques, less than 20, except in the fish which died. For them viremia reach confluency in cell culture prior their death. After this time, the viremia increased both in frequency and value until it reached 100% of fish. In the same time and with almost the same slope, the survival curve dropped. After day 120, the viremia readily decreased within a fortnight.

This heavy-mortality period corresponds to the temperature elevation from +7°C to +14 °C. This elevation seems to trigger the disease which had been present in latent condition until this time. But in fact, the viremia started climbing 60 days post infection, exceeding 80% of the fish by day 105 when the temperature had been ranging from +5 °C to +7 °C. So one cannot state whether or not the mortality would have been the same as that shown on the curve if the temperature had been maintained at its low values. Indeed, it is also possible that the temperature rise only accelerates the course of the disease.

On the other hand, the lack of susceptibility of fish at the beginning of the experiment, just after they had been living in summer-temperature water, could be substantiated by the fact that their protective mechanisms are still effective. But this efficacy is lowered by the wintering physiology and the infection can progress (i.e., after day 60).

At present, we raise the hypothesis that the carp must be much more susceptible to SVC after wintering than before wintering, when infected at the same temperature (i.e., 11 °C). New sets of experiments will be undertaken to solve these points.

The virus shedding into the water was detected between days 30 and 105; afterwards it disappeared, for which no explanation can be offered. It was detected much later than the gill infection. This accounts for the fact that the gill is not involved in the virus shedding.

Finally, the seroneutralization tests which had been performed on days 60 and 75 post infection provided negative results. However, one survivor displayed a positive neutralization 6 months post infection.

IP Infection Experiments (Fig. 2)

The IP infection trials appear confirmatory of the former results. The sinusoidal pattern of both gill infection and viremia curves is due to the consistent lower response of one of the two groups of infected carp. In the case of water route infection, the gill infection seems to be a reliable clue to SVC contamination. The viremia readily increases and the mortality is also more rapid. It is obvious that the entry of 5×10^3 pfu into a fish involves the infection of a sufficient number of target cells for accounting for the course of the disease.

The virus shedding was not investigated here. Indeed, we mainly aimed to observe the events following the contamination via the natural route of infection. However, the IP infection trials had been undertaken because former experiments had revealed some failures in reproducing the waterborne SVC. But since the gill biopsy indicated readily the present success of the water route infection trials, we focussed most of the technical work on them.

Conclusions

1) The SVC contamination occurring during autumn is followed by some clinical cases of SVC during winter and by massive mortalities when the temperature rises above 7 °C. During the most part of this period, the virus is shed into the water. So the introduction of a small number of autumn-contaminated carp into a virus-free fish stock would be enough to induce the subsequent occurrence of SVC. This autumn contamination can occur easily during the sorting operations when the fish are in a very close contact.
2) The previous results could be due to the wintering physiology acting either as a host defenses depressor or as a cell virus production depressor, and it may be that both mechanisms play a role
3) Technically speaking, the gill biopsy constitutes a reliable way of screening the fish for SVC contamination.

References

1. Fijan N, Petrinec Z, Sulimanovic D, Zwillenberg LO (1971) Isolation of viral causative agent from the acute form of Infectious Dropsy of Carp. Vet Arch 41:125–135
2. Kinkelin P de, Le Berre M (1974) Rhabdovirus des Poissons. II – Propriétés *in vitro* du virus de la Virémie Printanière de la Carpe, Ann Microbiol Inst Pasteur 125:113–124
3. Ahne W (1978) Uptake and multiplication of spring viremia of carp virus in carp, *Cyprinus carpio L.* J Fish Dis 1:265–268

Spring Viremia of Carp (SVC): Studies on Immunization of Carp

W. AHNE[1]

Abstract

One- and three-years-old carp (Cyprinus carpio) were infected by water route with spring viremia of carp virus (SVC). Infected carp kept at 10($\pm$1) °C, 15($\pm$1) °C, and 20($\pm$1) °C developed neutralizing antibodies 8, 7, and 5 weeks after infection. Circulating antibodies persisted during the period of experiment (17 weeks) and carp showed a protective immunity to a challenge (IP) with SVC virus.

Carp immunized by injection of SVC virus developed neutralizing antibodies within 28 days at 20 °C. Subsequent lowering of temperature of water (10 °C) and challenge with SVC virus did not influence the immune status of the carp.

Reference

Ahne W (1980) Rhabdovirus carpio-Infektion beim Karpfen (Cyprinus carpio): Untersuchungen über Reaktionen des Wirtsorganismus. Beih Zentralbl Veterinärmed 30:180–183

---

1 Institute for Zoology and Hydrobiology of the University of Munich, 8000 Munich, FRG

Studies on Immunization of Trout Against IPN

B. J. HILL[1], M. DORSON[2], and P. F. DIXON[1]

Introduction

For any vaccine to have practical value in the field, it must be safe (i.e., not produce the disease it is designed to protect against) and it must be effective (i.e., immunogenic enough to induce an effective level of protection). Little progress has been reported to date on the development of a practical vaccine against infectious pancreatic necrosis (IPN) of salmonids. It has been demonstrated (Dorson et al. 1975, 1978) that, after a relatively low number of serial passages through tissue culture, virulent IPN virus develops a sensitivity to neutralization by normal trout serum and at the same time becomes avirulent for trout fry. The neutralization of cell-culture-adapted (CCA) virus by normal trout serum has been shown to be due to an antibody-like non-virus-induced protein which has a sedimentation coefficient of 6S (Vestergard-Jørgensen, 1973; Dorson and de Kinkelin, 1974). It has been reported by Hill and Dixon (1977) that the acquisition of 6S sensitivity by IPN virus is not a permanent mutation or an adaptation process, but a simple selection of a fast-growing tissue culture variant from a mixed virus population, and that this process can be suppressed or even reversed by growth of the virus in the presence of normal trout serum.

Although avirulent, CCA virus has no value as a live vaccine, since rainbow trout fry infected with it are not protected against subsequent challenge with "wild" virulent virus of the same serotype (Dorson, 1977). Furthermore, other studies have shown that when in the 6S-sensitive form neither naturally virulent nor naturally avirulent IPN virus strains induce antibody production in adult trout after intraperitoneal inoculation, whereas when in their 6S-resistant form both naturally avirulent and naturally virulent strains, inoculated alive or inactivated, induce high levels of neutralizing antibody against virulent virus (Hill et al., unpublished results). The finding that trout antibodies raised against the naturally avirulent virus strain cross-reacted almost completely with the wild-type virulent virus suggested the possibility of the use of this strain as an avirulent vaccine for fry.

Studies have continued in an effort to induce effective immunization of young trout fry with inactivated or live vaccines and in this paper we report some of the main results obtained to date.

---

1 Ministry of Agriculture, Fisheries and Food, Directorate of Fisheries Research, Fish Diseases Laboratory, Weymouth, Dorset DT4 8UB, England
2 Institut National de la Recherche Agronomique, Laboratoire d'Ichtyopathologie, Thiverval-Grignon, France

Methods

Preparation of Viruses

Two strains of IPN virus (Sp serotype) have been used throughout: the French virulent strain 31/75 (Dorson et al., 1975) and the English avirulent strain described by Hill and Dixon (1977) and designed 74/53. The viruses were propageted in RTG-2 or BF cells in the presence of normal trout serum to ensure that they were in the 6S-resistant form. Virus preparations were used either as crude extracts of infected cells, or after concentration by 7% polyethylene glycol (PEG).

Inactivation of Virus

Crude or concentrated preparations of virus were treated with 1/200 formalin for 24 h at 4 °C, then dialysed against physiological saline for 48 h. Alternatively, inactivation was achieved by treatment with 1/250 β-propiolactone (BPL) at 37 °C for 3 h at pH 7.2.

Disruption of Virus

Concentrated virus preparations were disrupted by two different procedures known to release the constituent polypeptides from the virions. Procedure 1 involved treatment with a solution of 1% sodium dodecyl sulphate (SDS), 0.5 M urea and 10% acetic acid for 1 h at 37 °C. Procedure 2 involved treatment with 1% SDS, 1% 2-mercaptoethanol at 100 °C for 2 min. Following disruption, the preparations were dialyzed against distilled water for 24–48 h.

Fish

Young rainbow trout fry within the age range 4–8 weeks, obtained from known IPN-free commercial trout stocks, were used throughout.

Vaccination Methods

For intraperitoneal (IP) vaccination, fry were injected with 5–20 μl of vaccine preparation according to their size. In the hyperosmotic infiltration (HI) bath method, fry were immersed in a solution of 5% NaCl for 2 min prior to immersion in the vaccine solution for 10–20 min. For vaccination with live virus, fry were held in static water containing 10^4 plaque-forming units (pfu) of virus/ml for a period of 3–5 h, after which the normal water flow conditions were resumed.

Following vaccination, fry were held in flowing water at 10 °–12 °C with normal feeding for a period of two or three weeks before challenge.

Challenge

For all challenges, water flow to the fry tanks was stopped and suspensions of live virulent IPN virus strain 31/75 were added to give a final concentration of approximately 10^4 pfu/ml. After a period of 3–5 h, the water flow was resumed. Fish were observed daily and mortalities were recorded. Mortality figures quoted in the text represent the cumulative percentage mortality at 35 days post-challenge (%d_{35} CM).

Results and Discussion

Tests with BPL-Inactivated Virus

It has been reported (Dorson, 1977) that virulent IPN virus inactivated with formalin is immunogenic for rainbow trout fry. Treatment of virus with formalin, however, requires a prolonged dialysis period to remove toxicity prior to inoculation of fry. An alternative effective inactivant of many viruses is β-propiolactone, which has the additional advantage of self-hydrolysis to non-toxic residues after inactivation of the virus. Trial experiments established the effectiveness of BPL for inactivation of IPN virus. Moreover, intraperitoneal inoculation of adult trout with BPL-inactivated virus (without dialysis) resulted in the induction of high levels of neutralizing antibody against the virulent virus (Hill et al., unpublished results).

The immunogenicity of BPL-inactivated IPN virus strain 31/75 for rainbow trout fry was tested by intraperitoneal inoculation of 300 fry, followed two weeks later by challenge with virulent virus. The overall mortality of the vaccinated group was less than 20% of that of an equally challenged non-inoculated control group, thus confirming BPL-inactivated IPN virus to be immunogenic for trout fry.

Following this confirmation of immunogenicity, an attempt was made to immunize young trout fry by applying a BPL-inactivated crude preparation of virulent strain 31/75 by the HI bath method. The fry were challenged with virulent virus two weeks after "vaccination" and a mortality of 19% was obtained, compared to a mortality of 53% in the same period for a non-"vaccinated" control group, indicating that at least a moderate level of protection had been achieved. However, a subsequent repeat of this experiment failed to confirm that protection could be achieved with BPL-inactivated virus applied by the HI bath method.

The observed difference between the efficiency of IP inoculation and HI bath methods confirmed the findings for formalin-inactivated IPN virus vaccine reported by Dorson (1977). Since the immunogenicity of the inactivated vaccines had been clearly demonstrated by IP inoculation, it was concluded that the poorer protection induced by the HI bath method almost certainly resulted from an inefficient or inadequate uptake of the inactivated virus particles. It was considered that uptake of viral antigen might be improved if the virus particles were disrupted into smaller antigenic polypeptides.

Experiments with Disrupted Virus

Two main experiments with disrupted "vaccines" have been completed to date. In the first experiment a concentrated preparation of strain 31/75 was disrupted by procedure 1 (see Methods) and dialysed. A group of 150 four-week-old fry was inoculated IP and another 150 vaccinated by the HI bath method. A group of 150 fry was inoculated IP with formalin-inactivated virus (as a control on the immunocompetence of the fry) and a further 150 received formalin-treated non-infected RTG-2 cells (as a negative control for injected fish). Two other groups of fish were left untreated. All fish except for those in one of the non-treated groups were challenged with virulent virus three weeks later. The resulting cumulative mortality curves for each group are shown in Fig. 1.

The low mortality in the group inoculated IP with inactivated whole virus (7%) confirmed that the young trout fry were capable of being immunized against IPN.

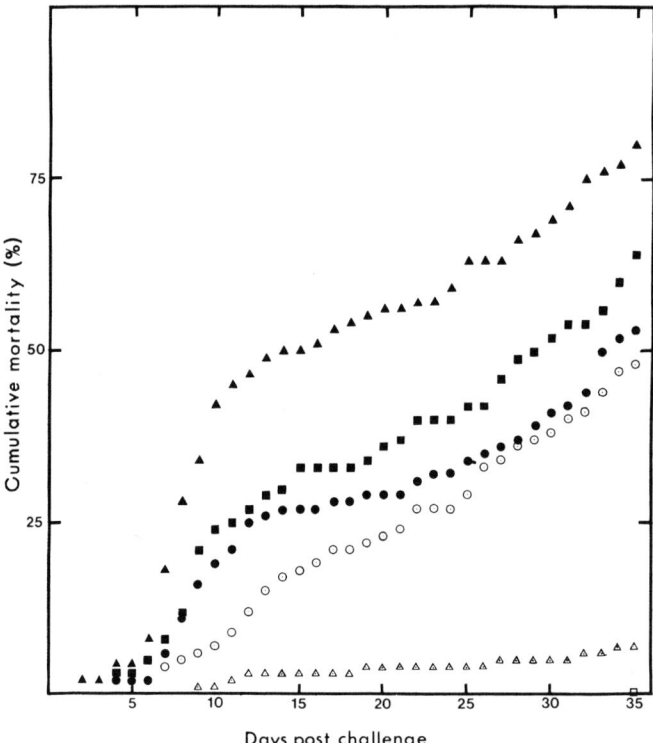

Fig. 1. Cumulative mortalities of rainbow trout fry challenged with IPN virus strain 31/75 three weeks post vaccination with inactivated or disrupted IPN virus.
▲ Non-vaccinated control fry
■ Fry intraperitoneally injected with disrupted virus
● Fry administered with disrupted virus by HI bath treatment
○ Fry intraperitoneally injected with formalin-treated RTG-2 cell extract
△ Fry intraperitoneally injected with formalin-inactivated virus
□ Non-challenged control fry

However, the hoped-for improvement in induced protection by use of disrupted virus with the HI bath application method was not realised (mortality 53%). The absence of significant protection in the group inoculated IP with disrupted virus (mortality 64%) indicated that a loss of immunogenicity of the viral proteins had occurred during the disruption procedure. It was not possible, therefore, to conclude whether disruption of the virus had led to an improved uptake of viral protein compared with whole virus when applied by the HI bath method.

In the second experiment a concentrated preparation of strain 31/75 was dissociated by procedure 2 (see Methods), and after dialysis used to treat duplicate groups of 300 four-week-old fry by the HI bath method. For comparison, two groups of 300 fry were treated with the non-disrupted concentrated virus after BPL inactivation or with a BPL-inactivated crude virus preparation, using the HI bath method. After two weeks, fish from one of each duplicate group were challenged with virulent strain 31/75 virus and the other group of each pair left unchallenged to test for total loss of infectivity of the vaccine. A group of 300 non-treated fry was also challenged at the same time, as a positive control for susceptibility of the fry to IPN and an assay of the virulence of the challenge dose. The results are shown in Table 1.

The results for the BPL-inactivated non-disrupted vaccine confirmed the earlier findings that such preparations applied by the HI bath method are not effective for protection against IPN. A slightly lower mortality for the fish vaccinated with dissociated virus could not be considered significantly different from the mortalities in the other groups and confirms the result of the previous experiment.

The antigenicity of the disrupted vaccine was tested in comparison with untreated virus in a double diffusion test against high-titre rabbit anti-IPN virus antiserum. The result is shown in Fig. 2.

The absence of the normal precipitin lines between antiserum and antigen confirmed the suspected loss of antigenicity in the disrupted virus preparation. Thus, it is not possible at this stage to conclude whether efficiency of uptake of viral antigen using the HI bath method of application could be improved by disrupting the viral

Table 1. Day 35 cumulative mortalities of rainbow trout fry challenged with IPN strain 31/75 two weeks after vaccination with inactivated or disrupted strain 31/75 administered by the HI method

Vaccine	Mortality of non-challenged vaccinated fry (d_{35} CM%)	Mortality after challenge (d_{35} CM%)
Non-vaccinated control	–	67
BPL-inactivated, crude virus	1	65
BPL-inactivated, concentrated virus	1	75
Disrupted, concentrated virus	1	58

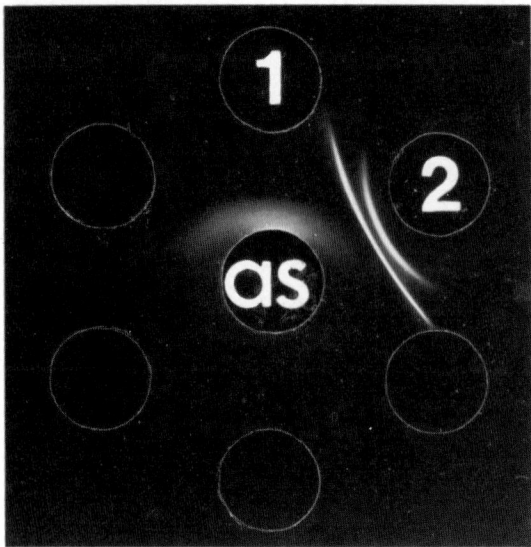

Fig. 2. Serological comparison of disrupted and untreated IPN virus by double diffusion in a 1% agarose gel in 0.9% sodium chloride. *1* IPN virus disrupted in 1% SDS, 1% 2-mercaptoethanol heated at 100 °C for 2 min; *2* untreated IPN virus; *as* IPN antiserum

particles. This question can only be settled when a means for disrupting the virus without loss of antigenicity of the viral protein is developed.

Live Virus Vaccines

In addition to the experiments with the inactivated and dissociated IPN virus vaccines, tests have been carried out on the possibility of using live naturally avirulent virus (strain 74/53) or live attenuated virus (strain 31/75). Independent pilot experiments with strain 74/53 in both laboratories provided encouraging results. In one experiment, fry vaccinated with the live avirulent virus by a simple bath infection method, and challenged two weeks later with virulent virus, reached an 11% cumulative mortality compared with 27% for the challenged non-vaccinated group. Likewise, in the other pilot experiment, fry vaccinated by simple bath infection with live virus and challenged two weeks later, reached a 9% cumulative mortality compared to 22% for the challenged non-vaccinated group. Additionally, in the second experiment the live vaccine was also administered to fry by the HI bath method and after similar challenge, a cumulative mortality of only 4% occured. This indicated the possibility that the HI method might be a more efficient means of applying a live avirulent vaccine than simple bath immersion. The HI method of application of the live virus was repeated on a larger scale, using groups of 300 six-week-old fry which were challenged two weeks after vaccination. The results of the pilot experiment were confirmed, with the vaccinated fry reaching a cumulative mortality of 18% compared with 53% for the non-vaccinated group.

Following these experiments, a direct comparison of the simple bath and the HI bath methods of delivering the vaccine was carried out to determine their relative efficacies. Four-week-old fry were vaccinated with the live avirulent virus by each of

Table 2. Day 35 cumulative mortalities of rainbow trout fry challenged with IPN strain 31/75 two weeks after vaccination with live avirulent strain 74/53 administered by two methods

Mode of administration of vaccine	Mortality of non-challenged vaccinated fry (d_{35} CM%)	Mortality after challenge (d_{35} CM%)
Non-vaccinated control	–	67
Bath method	5	55
HI method	4	32

the two delivery methods in duplicate groups of 300 fish each. Fish in one of each of the duplicate groups were challenged two weeks later by virulent virus. The non-challenged vaccinated fry groups served as controls on the mortality caused by the live vaccine itself or its method of administration. Non-vaccinated fry were also challenged as a positive control on the susceptibility of the fry and the virulence of the challenge virus. The results of the cumulative mortalities are shown in Table 2.

The fry infected by the simple bath method but not challenged subsequently showed only low mortalities (5%), confirming the avirulence of this vaccine virus. The protection given by the vaccine administered by the simple bath method was not as effective as suggested in the pilot experiments: 55% of the vaccinated fry died after challenge compared with 67% mortality in the challenged unvaccinated group. However, in contrast, fry vaccinated with the same virus preparation by the HI bath method had a moderate degree of protection, only 32% of fry succumbing to challenge. This result confirmed the previous finding that protection with live avirulent virus is enhanced when the application is by the HI bath method rather than by simple bath infection. The lower degree of protection by this method compared to the previous experiments might be a reflection of the differences in the age groups of the fry used in the two experiments. In the first experiments the fry were six weeks old at vaccination and were more completely protected than in the second experiments, when they were vaccinated at four weeks of age. The difference may reflect differing degrees of immunocompetence in the two age groups, although influence of genetic differences on susceptibility to IPN between the two groups of fry rather than an age difference cannot be ruled out.

The HI application method was used in a further experiment with live avirulent virus strain 74/53 to test the effectiveness of using lower doses of virus to infect the fish (10^2 pfu/ml rather than 10^4 pfu/ml). In contrast to the previous results, no protection whatsoever was achieved when the vaccine was diluted to 10^2 pfu/ml, suggesting that there may be a vaccine dose-effect on the degree of protection conferred by the live virus. Further experiments are planned to examine this dose-effect in more detail.

As an alternative to the use of the naturally avirulent strain 74/53, we have considered that the attenuated 31/75 strain may be a more effective immunogen. Attenuation of the virulent 31/75 strain has been attempted by long-term serial passage at 20 °C in the presence of normal trout serum. Unfortunately, a technical difficulty encountered with this work has been that yields of infectious virus have declined gradually with increased passage. In a single pilot experiment a 30% mortality resulted with the "attenuated" preparation as opposed to a mortality of 50% produced by the original virus under the same conditions. Serial passage of the virus is being continued in the hope of inducing a complete loss of virulence.

Other Tests

Other approaches for improving the efficacy of vaccination of fry against IPN are being explored at both laboratories. These include the use of a potent interferon inducer and use of a commercial vibrio vaccine (as an adjuvant) before and at the time of application of IPN vaccine preparations, in an attempt to enhance the immune response. Attempts are also being made to immunize fry at higher than normal temperatures (16 °C) before return to their normal hatchery temperature of 10–12 °C. Results to date have been inconclusive, but further studies are in hand.

References

Dorson M (1977) Vaccination trials of rainbow trout fry against infectious pancreatic necrosis. Bull Off Int Epiz 87:405

Dorson M, Kinkelin P de (1974) Nécrose pancréatique infectieuse des salmonidés: existence dans le sérum de truites indemnes d'une molécule 6S neutralisant spécifiquement le virus. C R Acad Sci Paris Ser D 278:785–788

Dorson M, Kinkelin P de, Torchy C (1975) Virus de la nécrose pancréatique infectieuse: acquisition de la sensibilité au facteur neutralisant du sérum de truite après passages successifs en culture cellulaire. C R Acad Sci Ser D 281:1435–1438

Dorson M, Castric J, Torchy C (1978) Infectious pancreatic necrosis virus of salmonids: biological and antigenic features of a pathogenic strain and of a non-pathogenic variant selected in RTG-2 cells. J Fish Dis 1:309–320

Hill BJ, Dixon PF (1977) Studies on IPN virulence and immunization. Bull Off Int Epiz 87 (5–6): 425–427

Vestergard-Jørgensen PE (1973) The nature and biological activity of IPN virus neutralizing antibodies in normal and immunized rainbow trout *(Salmo gairdneri)*. Arch Gesamte Virusforsch 42:9–20

Bacterial Diseases

Chairman: A. L. S. MUNRO, Scotland

Experimental Vibriosis in the Eel (Anguilla anguilla)

H. CHART and C.B. MUNN[1]

Introduction

Vibriosis is probably the most important disease of marine and, to a less extent, freshwater fishes. With an increase in fish aquaculture the disease has become a severe economic problem.

The organism was first mentioned by Canestrini in 1893 [5], who named the bacterium responsible for an epizootic in eels: *Bacillus anguillarum,* later to be renamed *Vibrio anguillarum* by Bergman in 1909 [3]. In eels the first external symptoms consisted of reddening of the tail and fins, and inactivity. Post-mortem examination revealed congested intestinal blood vessels, liver, and genitalia [4]. Bergman [3] reported the presence of bloody lesions in the musculature of infected fish. Literature on non-anguillid fish is too extensive to outline in this paper, however one particular feature of the disease is the presence of fluid in the alimentary canal variously described as a bloody discharge from the vent [15], clear viscous fluid in the gut [2, 11] and mucus in the gut [20]. Although symptoms of vibriosis in fish are well described, mechanisms of pathogenicity of *V. anguillarum* and the events leading to the invasion and death of the fish host appear to be unknown. Several theories for the route of entry of the pathogen have been put forward, the main ones being as a contaminant in fish-based diets [19] or via contaminated water [18, 19].

The aims of this present investigation were to determine the effects of administering inocula of *V. anguillarum* via different routes into the freshwater eel *(Anguilla anguilla),* and the nature of a mucoid, anal exudate encountered during the experiment.

Material and Methods

Inoculation of Fish

Strains of *V. anguillarum* were obtained from various sources isolated from naturally infected fish. For virulence testing, inocula of 0.1 cm^3 (10^7 organisms) were injected intraperitoneally (IP) into groups of eels (50–70 g wt) kept in 50% sea water. Fish were also inoculated with virulent strains by intramuscular (IM) or intravascular (IV) injection, or by insertion into either the foregut or rectum by means of a catheter. Control fish received 0.1 cm^3 sterile Tryptone soy broth (TSB). Fish were externally

---

1 School of Environmental Sciences, Plymouth Polytechnic, Plymouth, PL 4 8AA, UK

challenged by placing them in 50% sea water containing approximately 10^6 viable organisms cm^{-3}. This was carried out on both intact fish and eels with a 1-cm lesion in their flank.

Culture of Organisms

Test strains were cultured in TSB [(Oxoid) 1.5% NaCl, 25 °C, 24 h]. Immediately post mortem, blood, spleen and anal exudate samples were streaked onto blood agar plates. Resulting colonies were cultured in TSB; motile, gram-negative, curved rod-shaped bacteria which were fermentative, oxidase-positive and sensitive to the vibriostatic compound 0/129 (2,4 diamino-6,7 diisopropyl pteridine phosphate) were considered *Vibrio* species. To confirm Koch's postulates, eels were injected with these isolates.

Histology

Tissues to be examined by light microscopy were fixed in 10% Formol-saline for two days. After decalcification in R.D.C. (Bethlehem Instruments Ltd.), material was dehydrated in graded alcohols, cleared in xylene and embedded in paraffin wax. Sections (10 µm) were cut using a rotary microtome, and stained with Mallory's triple stain and periodic acid Schiff's stain (PAS) for mucopolysaccharides [17].

Tissues to be examined by transmission electron microscopy (Phillips 300 TEM) were fixed in 3% gluteraldehyde in saline, post-fixed in 1% osmium, dehydrated in graded alcohols and treated with propylene oxide prior to resin embedding. Ultrathin sections cut on a Porter-Blum Ultramicrotome were stained with uranyl acetate and Reynold's lead citrate. Materials examined by scanning electron microscopy (Jeol 35C SEM) were fixed in 3% gluteraldehyde in saline (1 week) and osmicated (2 days). After dehydration the tissues were critical-point-dried (Sam Dry PVT 3) and gold-coated.

Results

Effects of Intraperitoneal Injection

From a total of 15 test strains injected IP, 7 were avirulent and 8 virulent. Virulent strains were designated as those causing death within 2 days. In most cases, the symptoms consisted of severe haemorrhaging, ulceration and liquefaction of the musculature at the site of injection. Petechial haemorrhages were present on the ventral surfaces, including the lower jaw. Congestion was prevalent in the fins and the gills were haemorrhagic. Internally, petechial haemorrhages were abundant on the wall of the peritoneal cavity and the entire alimentary canal was haemorrhagic with enlarged, congested blood vessels. The kidney, spleen and liver were haemorrhagic in most cases but the heart appeared unaffected. In most cases, anal exudate was present, but not all had fluid in the gastric caecum.

Effects of Inoculation by Other Routes

To determine the effect of various inoculation routes, fish were injected with a virulent strain. In all cases fish injected IP, IV, IM died within 2 days of inoculation. Eels infected via the rectum died after 15–16 days whilst orally challenged individuals did not succumb to infection within 3 weeks. All fish maintained in water containing viable vibrios survived longer than 3 weeks. Eels injected IM and IV had very large purple-red ulcerous lesions containing liquefied tissue at the injection site, and reddening of the gut. Necrosis of the organs, gills and the presence of anal exudate was common to both routes. Eels infected via the rectum showed symptoms including haemorrhaging and congestion of the fins and skin. Internally, the spleen, liver and kidney were haemorrhagic with reddening of the hindgut. There was not the severe haemorrhaging and congestion encountered in the injected routes, although anal exudate was present.

Histology

The results of electron microscopy on the anal exudate revealed that this was predominantly portions of the gut epithelium (Fig. 1). Examination of infected and control wax sections, taken at intervals along the gut, showed that the epithelia were affected along the entire length of the alimentary canal except the oesophagus (Fig. 2). Sections stained with PAS demonstrated that epithelia were detaching at the basement membrane of the gut. This was supported by SEM which showed that the lamina propria remained intact (Fig. 3).

Bacteriology

V. anguillarum could be isolated in pure culture from both blood and spleen of infected fish. The anal exudate contained large numbers of *V. anguillarum* along with commensal gut bacteria. In all cases, bacteria designated *Vibrio* species fulfilled Koch's postulates.

Discussion

Although some strains were deduced avirulent, these may become virulent under different conditions. Relationships between the experimental routes and the natural route(s) of entry are unknown. However, it was evident that once a virulent strain gained access into the fish tissues, disease and eventual death was imminent. Disease symptoms and death resulted in eels receiving anal inocula, although it is unknown whether this is an important route of entry. However, it demonstrated that the organism could remain viable in the rectum and eventually invade the tissues. It may be that a variant of the organism has to be selected or that a minimum population size has to be reached. The oral route of inoculation was of particular interest as this has been proposed by several authors as of importance in natural infection [18, 19], but in the present investigation no deaths occurred within 3 weeks in fish receiving large

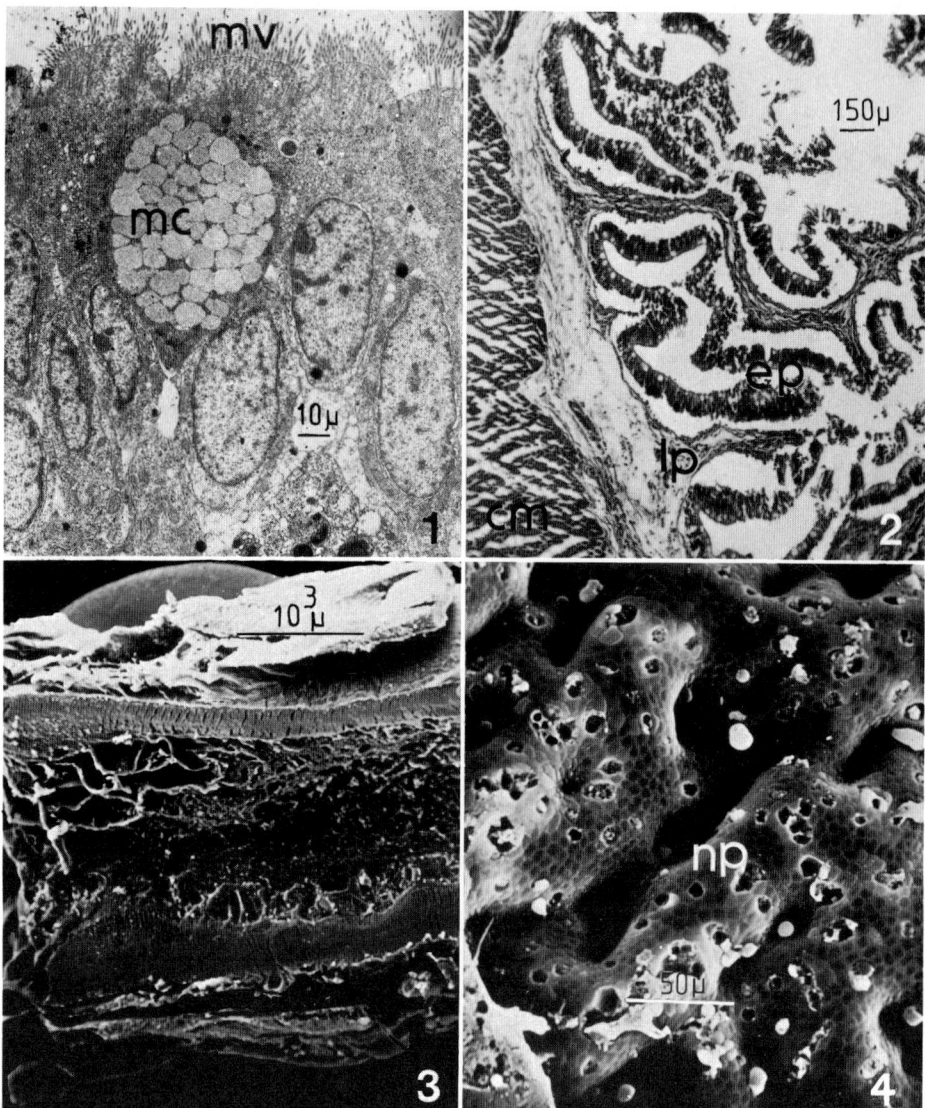

Fig. 1. Electron micrograph of gut epithelium from anal exudate *mv* microvilli; *mc* mucus cell

Fig. 2. Light micrograph of infected gut (Mallory's triple stain) *cm* circular muscle; *lp* lamina propria; *ep* epithelium

Fig. 3. Scanning electron micrograph of gut

Fig. 4. Scanning electron micrograph of gastric caecum *Np* Necrotic pits

inocula by this route. It was very surprising that fish with or without lesions, exposed to water containing vibrios, did not contract vibriosis in the three-week period, since vibrios have been found to invade host tissues by means of dermal lesions inflicted by tag labelling [18] and sea louse attachment [10].

It was evident that, regardless of inoculation route, in most diseased fish the gut epithelium had broken down, constituting a large proportion of the anal exudate. Unfortunately, there is very little information available on the effects of disease on fish tissues. Consequently, data published on mammalian pathology had to be consulted to draw parallels between situations occurring in fish and known to occur in mammals. It appears that in this case a type of inflammation is taking place. The frequently observed vasodilation of normally minute gut vessels appears to be analogous to arterial hyperaemia which takes place during inflammation. Inflammation is known to occur in fish and is closely comparable to that of mammals [8]. The irreversible cell necrosis occurring in the gut of fish with vibriosis has strong resemblance to suppuration and catarrhal inflammation exhibited in mammalian pathology [12]. Although it seems highly unlikely that normal host tissue would be affected detrimentally by its own leukocytes [12], it is possible that the presence of *V. anguillarum* or the products of necrosis caused by the action of toxins on host tissues could act as chemotactic agents for leukocytes. There have been several reports of *V. anguillarum* having gelatinase enzymes [9, 14, 16] which may play a part in the lysing of host tissues. *V. anguillarum* produces a haemolysin [13] which may be cytolytic to cells other than erythrocytes [1].

The involvement of leukocytes in the epithelial necrosis is not clear but, in the anal exudate debris, cells which may be leukocytes were present. Unfortunately, no literature on eel leukocytes has been encountered and there seems to be no clear correlation with leukocytes of other fish species [6, 7]. The steps which lead to the eventual detachment of the gut epithelium have not been investigated; however the gastric caecum from an eel that had died of the disease exhibited what would seem to be early stages in necrosis of the epithelia (Fig. 4). This focal necrosis would seem unlikely to occur in the hindgut region as sheets of epithelium are dislodged.

When studying the gut epithelia of eels the following must be considered. Firstly, eels are known to alter the structure of the gut during metamorphosis from the yellow to silver conditions. Secondly, when eels are transferred to different salinities, adaptation to the new environment involves the oesophageal epithelium [22]. The effect this has on the remainder of the alimentary canal is not known.

The reasons for the phenomena outlined in this paper and their importance in the ultimate death of fish are not clear, and further investigation is in progress.

Acknowledgement. H.C. would like to thank the SRC for the post-graduate studentship under which this work was carried out. These experiments were licenced by Home Office Licence No. SWI 2789.

References

1. Alouf JE (1977) Cell membranes and cytolytic bacterial toxins. In: Cuatrecasas P (ed) The specificity and action of animal, bacterial, and plant toxins, vol. I. Chapman & Hall, London, p 338
2. Anderson JIW, Conroy DA (1970) Vibrio disease in marine fishes. In: Sniezko SF (ed) A symposium on diseases of fishes and shell fishes. American Fisheries Society, Washington, pp 266–272
3. Bergman AM (1909) Die rote Bealenkrankheit des Aals. Ber Bayer Biol Versuchsanst, München 2:10–54
4. Brunn A, Heiberg B (1932) The red disease of the eel in Danish waters. Medd Komm Havundersøg KBH Ser Fish 9:1–19
5. Canestrini G (1893) La malattia dominante delle anguille. Atti Ist Veneto 7:809–814
6. Ellis AE (1976) Leucocytes and related cells in the plaice *Pleuronectes platessa*. J Fish Biol 8:143–156
7. Ferguson HW (1976) The ultrastructure of plaice *(Pleuronectes platessa)* leucocytes. J Fish Biol 8:139–142
8. Finn JP, Nielson NO (1971) The inflammatory response of rainbow trout. J Fish Biol 3:463–478
9. Håstein T, Holt G (1972) The occurrence of vibrio disease in wild Norwegian fish. J Fish Biol 4:33–37
10. Håstein T, Bergsjo T (1976) The salmon lice *(Lepeopheirus salmonis)* as the cause of disease in farmed salmonids. Riv Ital Pisic Ittiopat 11:3–5
11. Harbell SC (1976) The pathology of vibriosis in coho salmon. MSc Thesis, Washington
12. Muir R (1964) Muir's textbook of pathology, 8th edn. Arnold, London
13. Munn CB (1978) Haemolysin production by *Vibrio anguillarum*. FEMS Microbiol Lett 3:265–268
14. McCarthy DH (1974) Vibriosis in rainbow trout. J Wildl Dis 10:2–7
15. Novotny AJ (1975) Vibriosis: A common disease of Pacific salmon cultured in marine waters of Washington. Ext Bull Wash State Univ Coop Ext Serv (No. 663) p 8
16. Pacha RE (1969) Characterization and relatedness of marine vibrios pathogenic to fish; physiology, serology and epidermiology. J Bacteriol 100:1242–1247
17. Pearse AGE (1968) Histochemistry. Theoretical and applied, 3rd edn. J & A Churchill Ltd., London
18. Roberts RJ (1973) The histopathology of salmon tagging, III. Secondary infections associated with tagging. J Fish Biol 5:621–623
19. Ross AJ, Martin JE, Bressler V (1968) *Vibrio anguillarum* from an epizootic in rainbow trout *(Salmo gairdneri)* in the USA. Bull Off Int Epiz 69:1139–1148
20. Rucker PR (1956) Vibrio infections among marine and freshwater fish. Prog Fish Cult 21:22–25
21. Willis RA, Willis AT (1972) Principles of pathology and bacteriology, 2nd edn. Butterworths, London
22. Yamamoto M, Hirano T (1978) Morphological changes in the oesophageal epithelium of the eel *Anguilla japonica* during adaptation to sea water. Cell Tissue Res 192:25–38

Laboratory and Field Observations on Antivibriosis Vaccines

T.P.T. EVELYN and J.E. KETCHESON[1]

Introduction

Attempts in British Columbia at rearing Pacific salmon in floating netpens in sea water were first undertaken on an experimental (pilot-scale) basis at the Pacific Biological Station, Nanaimo, in 1974. Objectives of the study were to determine just how rapidly market-size salmon could be produced and to determine whether all of the five indigenous species of Pacific salmon were suitable mariculture candidates. After only one year of operation, it appeared obvious that mariculture ventures would likely only be successful if effective control measures for infectious diseases such as furunculosis, bacterial kidney disease, and vibriosis could be developed. Consequently, studies to examine the feasibility of using vaccination to control these diseases were initiated.

The purpose of this paper is to report our findings with respect to antivibriosis vaccines. Our experience had shown vibriosis to be a particularly acute problem, and, as with the US findings in Puget Sound, Washington (Harrell et al., 1976; Harrell, 1978), to be due to two serologically and biochemically distinct vibrios. In this paper, we refer to these vibrios as *Vibrio anguillarum* and *Vibrio* sp.

Materials and Methods

Bacteria

Two vibrios were used for vaccinating and challenging fish: *V. anguillarum,* strain V1 (Evelyn 1971), and *Vibrio* sp., strain 74/48. The latter was isolated in 1974 from a seawater-reared sockeye salmon *(Oncorhynchus nerka)* that had died of vibriosis. The vibrios were the biochemical and serological equivalents of the Puget Sound isolates 775 and 1669, respectively (see Harrell et al., 1976), and their virulence was maintained by passage in fish and storage at $-90\ °C$.

Fish and Fish Handling

Test fish were reared from eggs obtained from stocks of British Columbia salmon. The sockeye were derived from the Fulton River stock and the coho *(O. kisutch)* and chinook *(O. tshawytscha)* from Big Qualicum River stocks. During handling for

---

[1] Department of Fisheries and Oceans, Pacific Biological Station, Nanaimo, B.C., Canada

tagging, vaccination, and challenge, the fish were anesthetized with 2-phenoxyethanol (1/7,000). Tagging to denote the variously vaccinated and control fish was necessary because following (or during) challenge, fish in any given experiment were held in a common tank or netpen to ensure that all treatments experienced the same environmental conditions. Fish in field experiments received coded wire nose tags; fish in short-term laboratory experiments were appropriately fin-clipped.

Vaccines and Vaccination Methods

Cultures serving as the source of vaccines were grown at 21 °–24 °C in aerated brain heart infusion (BHI) broth (Difco) for 30 to 64 h. The cultures were then all killed by the addition of formalin (0.3% V/V).

Vaccines took several forms and were administered by injection, feeding, or immersion. Injected vaccines (intact cells, spent culture fluids) were given in a volume of 0.1 ml and as a single intraperitoneal (IP) injection; in some cases, the vaccines also contained a Freund's-type adjuvant. Fed vaccines consisted of intact cells; the cells in saline were sprayed onto the partially dried fish diet (Oregon Moist Pellets), and the latter was then fed to the fish for 14 or 15 days. Immersion-administered vaccines contained intact cells, disrupted cells, whole broth cultures, or spent culture medium (broth). They were administered to fish by a two-step hyperosmotic infiltration procedure very similar to that described by Croy and Amend (1977). The procedure involved a 2-min dip in oxygenated brine (7.1% NaCl and 0.9% Hanks' balanced salt mixture; pH 7.0) followed by a 3-min dip in oxygenated vaccine. The brine and vaccine were used at or below ambient water temperature, and Antifoam B (Dow Corning Corp.) was added, as necessary, to suppress foaming.

Dosages of bacterial cells used in the vaccines (see Tables 1, 2, 3, 6, 7) are expressed in terms of wet weight. When disrupted cells were administered, the vaccine contained both the soluble and insoluble cell components. Cell disruption was accomplished with glass beads in a cell homogenizer (Bronwill Scientific, Inc.). When spent broth was tested as an immunogen, it was first freed of cells by centrifugation and filtration (0.45 μm pore diameter filter, Millipore Corp.). Further fractionation of the broth was achieved by ultrafiltration using an XM100A membrane (Amicon Corp.). The ultrafiltered fractions were kindly provided by Dr. R. Lallier, University of Montreal, Faculty of Veterinary Medicine.

Following vaccination, fish in the first of two field experiments were held in freshwater at 6 °C for 8 weeks and at 16 °C for 2 weeks until challenged. Vaccinated fish in the second field experiment were held in freshwater at 12 °–13 °C for a month until challenged. In the laboratory experiments, vaccinated fish were, with one exception (Table 3), held in water at 12 °–15.5 °C until challenged.

Challenge Methods

The efficacy of vaccination was determined by natural and experimental challenge. Natural challenge was used in two field experiments (Tables 1 and 2); it occurred when fish were transferred to sea water during warmwater periods.

Experimental challenge was accomplished by IP injection (0.1 ml). It was usually effected by injecting vaccinated and control fish with a dose of vibrio cells sufficient to kill 70%–85% of the controls. In a few cases, the efficacy of vaccination was assessed by determining and comparing the doses (LD_{50} values) of injected vibrio cells required to kill 50% of the vaccinated and control fish. LD_{50} values were calculated by the method of Reed and Muench (1938).

All mortalities resulting from natural and experimental challenge were cultured to determine the cause of death and examined to identify the treatment received. In addition, fish receiving experimental challenge were held in water at 13°–15.5 °C to allow the challenge to express itself. Experimental challenge experiments were terminated only when mortalities had ceased for 3 or 4 days.

Results and Discussion

Tables 1 and 2 give the results of two experiments initiated in 1975 and 1976, respectively, to investigate, in part, the efficacy of vaccination against vibriosis under conditions of natural challenge in sea water.

In the first experiment, sockeye fingerlings, vaccinated as outlined in Table 1, were transferred to sea water on August 21. Losses due to vibriosis occurred for 14 weeks. The major challenge came from *Vibrio* sp., which killed four times as many control fish as *V. anguillarum*. (Challenge from *Aeromonas salmonicida* did not occur.)

The results (Table 1) clearly demonstrate the value of vaccination. All vaccines provided significant protection (P=<0.001) against vibriosis, the oral vaccine being decidedly less protective than its injected trivalent counterpart (P=<0.001).

The univalent *V. anguillarum* vaccine protected strongly against homologous challenge and also resulted in some cross-protection against *Vibrio* sp. (mortalities due

Table 1. Mortalities of variously vaccinated 6.5-g sockeye salmon during their first 14 weeks of natural challenge in sea water[a]

Vaccines[b]: method of administration		No. fish dead/ No. treated	Percent mortality due to:		Percent treated fish dead of:	
			All causes	Vibriosis	V a	Vx
None	–	260/455	57.1	56.5	11.4	45.0
Va:	IP	68/294	23.1	22.8	0.3	22.4
Vx:	IP	11/382	2.9	0.5	0.5	0.0
VxVa	IP	9/395	2.3	0.8	0.3	0.5
VxAs:	IP	17/387	4.4	2.1	1.5	0.5
VxVaAs:	IP	11/451	2.4	0.7	0.2	0.4
VxVaAs:	oral	61/460	14.6	10.7	5.4	5.2

[a] Water temperatures at 8 m rose from 10° to 14.5 °C at the start of the epizootic and fell to 7.5 °C by the end of the epizootic

[b] Each killed pathogen (Va, *V. anguillarum*: Vx, *Vibrio* sp: As, *A. salmonicida*) was injected in saline at a rate of 3 mg/fish or fed at a rate of 11.5 mg/fish over a 15-day period)

Table 2. Mortalities due to vibriosis among variously vaccinated 12.5-g sockeye salmon during their first 12 weeks of natural challenge in sea water[a]

Vaccine treatment[b]	No. fish treated	Percent fish dead of: Va	Vx
None (control)	383	42.0	0
Immersion method:[c]			
2 dips in vaccine, 14 days apart	384	0.8	0
1 dip in vaccine	367	1.9	0
1 dip in vaccine diluted 10^{-1}	378	4.2	0
1 dip in vaccine diluted 10^{-2}	383	5.0	0
Single IP injection:[d]			
Vaccine in saline	415	0.7	0
Vaccine in adjuvant	421	0	0
Vaccine (diluted 10^{-1}) in adjuvant	465	0.2	0
Vaccine (diluted 10^{-2}) in adjuvant	450	1.1	0
Oral[e]			
Vaccine fed for 14 days	393	17.6	0

[a] Water temperatures at 8 m were $11.5°-12°C$ during epizootic
[b] All vaccines contained cells (interact or disrupted) of 4 fish pathogens: *V. anguillarum*, *Vibrio* sp., *A. salmonicida*, and the kidney disease bacterium
[c] The undiluted immersion vaccine contained 3.5 g of disrupted cells of each pathogen/liter of 0.1% NaCl. When the vaccine was diluted, the diluent was 0.1% NaCl
[d] Each fish was injected with 2.0 mg (or less as indicated) of cells of each pathogen suspended in saline or a Freund's-type adjuvant
[e] Each fish received a total of 14.0 mg of cells of each pathogen

to *Vibrio* sp. were reduced by 50%). Such cross-protection has recently been reported by Schiewe and Hodgins (1977) and by Harrell (1978). More interesting were the results obtained with the univalent *Vibrio* sp. vaccine. This vaccine gave essentially complete protection against both homologous and heterologous challenge, thus raising the possibility that it could be used in place of a bivalent vibrio vaccine. Cross-protection by *Vibrio* sp. against *V. anguillarum* has been demonstrated by Schiewe and Hodgins (1977), but the protection was only partial. The different findings may be due to the fact that we used an unheated vaccine, or may simply reflect an inadequate challenge by *V. anguillarum* in the present experiment. Until this matter can be resolved the replacement of a bivalent vibrio vaccine with the univalent *Vibrio* sp. vaccine would be unjustified.

Table 2 summarizes the results obtained in a second field experiment in which variously vaccinated sockeye fingerlings were transferred to sea water in July. On this occasion, natural challenge came only from *V. anguillarum*, but the challenge was substantial. Once again, significant protection ($P < 0.001$) was provided by all vaccine treatments, with the oral vaccine again showing the least efficacy.

The immunogenicity of the vibrio immunogens was impressive. Injected vaccines containing vibrio cells in amounts as low as 20 µg/vibrio (4 µg dry weight/vibrio) re-

Table 3. Relative immunity at 7 months post vaccination of sockeye salmon[a] vaccinated against *Vibrio* sp.

Vaccination method[b]	LD_{50}[c]	Relative increase in immunity (LD_{50} control-LD_{50} treated)
None (control)	$10^{-7.0}$	–
IP injection	$10^{-2.0}$	$10^{5.0}$-fold
Immersion	$10^{-3.6}$	$10^{3.4}$-fold
Oral	$10^{-4.6}$	$10^{2.4}$-fold

[a] Fish averaged 12.5 g at vaccination and were held, post vaccination, in freshwater averaging 10 °C (temperatures gradually fell from 17° to 4 °C)

[b] Fish received 2 mg of cells by injection and 14 mg of cells by feeding; the vaccine used in the immersion method contained 3.5 g of disrupted cells/liter of 0.1% NaCl

[c] LD_{50} values represent the dilution of a *Vibrio* sp. challenge cell suspension required to kill 50% of the test fish. Challenged fish were held in freshwater at 14 °C

sulted in virtually full protection (mortalities of only 1.1%). This high degree of immunogenicity was also evident with the immersion vaccine, which, following only one application, provided significant protection even when diluted 100-fold. The ability to effect immunity using dilute vaccines (see also Croy and Amend, 1977) could reduce the costs of vaccinating fish against vibriosis.

The strength and duration of protection elicited by the various vaccination methods was briefly investigated using the LD_{50} approach and *Vibrio* sp. (Table 3). The data support our field results which suggested that injection and feeding, respectively, confer the highest and lowest levels of immunity. They differ in this respect from the field findings of Antipa and Amend (1977) which indicated immersion to be more protective than injection. The data also suggest that antivibriosis immunity is quite durable. The durability may have been enhanced somewhat by the low water temperatures (4°–7 °C) prevailing during the last 8 weeks of holding, but it could not have been affected by fish-pathogen contact because contact was precluded by holding the fish in freshwater.

Concerns regarding the duration of antivibrio immunity in sea water are probably unnecessary because natural challenge with vibrios probably serves to elicit or perpetuate immunity. It has been shown that contact with living vibrios stimulates immunity (Braaten and Hodgins, 1976; Egidius and Andersen, 1979), and this phenomenon is probably the major reason for the increased immunity displayed by our unvaccinated (control) fish during their second summer in sea water (Table 4).

The value of immersion vaccination as a means of mass immunization against vibriosis was considerably increased when it was found that very small fish could be effectively immunized. Also, vaccine preparation was greatly simplified when tests showed that vaccines consisting of disrupted cells could be replaced by killed broth cultures. Early tests on the foregoing in our laboratory utilized 2.2 g chinook salmon that had been dip-vaccinated in a killed broth culture of *Vibrio* sp. At 17 days post vaccination, these fish were found by the LD_{50} challenge method to be over 100 times as resistant to homologous challenge as their unvaccinated counterparts. Strong

Table 4. Mortalities due to vibriosis among unvaccinated (control) sockeye salmon during their first and second summers in sea water

Sockeye[a] (Experiment No.)	No. of summers in sea water	Vibrio challenge No.	Vibrio challenge Date	Percent fish dead (no. dead/no. challenged) of vibriosis
1	1	1	Aug. 26–Dec. 9, 1975	56.5 (257/455)
	2	2	Sept. 1–Oct. 14, 1976	1.2 (1/84)
2	1	1	Sept. 1–Oct. 14, 1976	40.0 (157/374)
	2	2	July 6–Aug. 18, 1977	1.5 (1/66)
3[b]	1	1	July 6–Aug. 18, 1977	19.3 (29/150)
	2		–	–

[a] Sockeye in the various experiments were held in adjacent netpens
[b] Observations on this group of fish had to be prematurely terminated during a period of vibrio challenge because of a sudden bloom of the phytoplankton *Chaetoceros convolutus*, which killed virtually all of the fish

Table 5. The immune response following immersion (dip-)vaccination with *Vibrio* sp. of 1.5–2.0-g coho salmon held in diluted (20‰) sea water and in freshwater at 15 °C

Holding water	Treatment[a]	No. of fish treated	Percent killed by *Vibrio* sp. challenge[b]
20‰ sea water	Dip-vaccinated	73	0[c]
Freshwater	Dip-vaccinated	83	0[c]
Freshwater	Unvaccinated	78	85.5

[a] The immersion vaccine consisted of a killed 48-h *Vibrio* sp. BHI broth culture, diluted 1/2 with 0.1% NaCl
[b] At 21 days post vaccination, all fish were converted to 20‰ sea water at 15 °C and challenged by an IP injection of *Vibrio* sp. (1.25×10^5 cells/fish)
[c] Chi square test: P=<0.001

immunity also resulted when 1.5–2.0 g coho were similarly vaccinated, even though some of the coho may have been under considerable stress because of premature holding in 20‰ sea water (Table 5).

Although we have not attempted to isolate or identify the vibrio immunogens, it would probably be instructive to do so since it might then be possible to learn something about the route and mechanism of their uptake. We know, however, that the immunogens must be soluble compounds because they occur in spent broth (Table 6). They are probably also slowly released from intact cells, because even well-washed cells are immunogenic when administered by dipping (Table 6). Finally, the immunogens must be large molecules (over 100,000 MW) because they are retained by XM100A ultrafilters (Table 7). These properties, listed for the immunogens of *V. anguillarum* in Tables 6 and 7, are also shared by the immunogens of *Vibrio* sp. Considering the large size of the immunogen molecules, it hardly seems likely that simple diffusion ex-

Table 6. Tests on the immunogenicity of intact cells and spent culture fluid (broth) derived from a killed 48-h BHI broth culture of *V. anguillarum*

Vaccine[a]	Method of administration	No. of fish[b] treated	Percent killed by *V. anguillarum* challenge[c]
Intact cells	IP injection	20	0[d]
Spent broth	IP injection	20	0[d]
Intact cells	Immersion	20	5[d]
Spent broth	Immersion	20	5[d]
None	–	20	70

[a] Intact cells were 3x saline-washed; they were injected (2 mg/fish) or used at 4 g/liter of 0.85 NaCl as an immersion vaccine. Injected vaccines contained a Freund's-type adjuvant
[b] Coho salmon averaging 10 g were used; they held in freshwater at 13°–14°C throughout the experiment
[c] Fish were challenged at 35 days post vaccination by injection (1.5×10^6 *V. anguillarum* cells/fish)
[d] Chi square test: P=<0.001

Table 7. Tests on the immunogenicity of two culture fluid (broth) fractions derived from a killed 64-h BHI broth culture of *V. anguillarum*

Vaccine[a]	No. of fish treated[b]	Percent killed by *V. anguillarum* challenge[c]
XM100A broth filtrate	20	80
XM100A broth retentate	20	25[d]
Killed *V. anguillarum* cells	20	30[d]
Saline	20	80

[a] Vaccines contained 0.1 ml of filtrate, retentate, or saline, or 2 mg of killed *V. anguillarum* cells, in a Freund's-type adjuvant; they were administered by injection
[b] Coho salmon averaging 25 g were used; they were held in freshwater at 15.5°C throughout the experiment
[c] Fish were challenged 30 days post vaccination by injection (~1.5×10^8 *V. anguillarum* cells/fish)
[d] Chi square test: P=<0.001 for broth retentate: P=<0.01 for killed *V. anguillarum* cells

plains their entry into fish. It seems more probable that the immunogens attach to certain sites on the fish and thereafter undergo some form of active uptake.

References

Antipa R, Amend DF (1977) Immunization of Pacific salmon: comparison of intraperitoneal injection and hyperosmotic infiltration of *Vibrio anguillarum* and *Aeromonas salmonicida* bacterins. J Fish Res Board Can 34:203–208

Braaten BA, Hodgins HO (1976) Protection of steelhead trout *(Salmo gairdneri)* against vibriosis with a living low-virulence strain of *Vibrio anguillarum*. J Fish Res Board Can 33:845–847

Croy TR, Amend DF (1977) Immunization of sockeye salmon *(Oncorhynchus nerka)* against vibriosis using the hyperosmotic infiltration technique. Aquaculture 12:317–325

Egidius EC, Andersen K (1979) Bath immunization — a practical and non-stressing method of vaccinating sea farmed rainbow trout *Salmo gairdneri* Richardson against vibriosis. J Fish Dis 2:405–410

Evelyn TPT (1971) First records of vibriosis in Pacific salmon cultured in Canada, and taxonomic status of the responsible bacterium, *Vibrio anguillarum*. J Fish Res Board Can 28:517–525

Harrell LW (1978) Vibriosis and current salmon vaccination procedures in Puget Sound, Washington. Mar Fish Rev 40:24–25

Harrell LW, Novotny AJ, Schiewe MH, Hodgins HO (1976) Isolation and description of two vibrios pathogenic to Pacific salmon in Puget Sound, Washington. Fish Bull US 74:447–449

Reed LJ, Muench H (1938) A simple method of estimating fifty percent end points. Am J Hyg 27:493–497

Schiewe MH, Hodgins HO (1977) Specificity of protection induced in coho salmon *(Oncorhynchus kisutch)* by heat-treated components of two pathogenic vibrios. J Fish Res Board Can 34:1026–1028

Recent Experience of Field Vaccination Trials Against Vibriosis in Rainbow Trout (Salmo gairdneri)

T. HÅSTEIN[1], F. HALLINGSTAD[2], T. REFSTI[3], and S.O. ROALD[1]

Introduction

Vibriosis is one of the most serious infectious fish diseases occurring in sea water, and the disease is often the major problem in the rearing of salmonid fish, causing considerable economic losses in the marine fish farming industry. These losses are due to increased mortality, reduced growth, and costs of medical treatment.

Since medical treatment is expensive as well as hazardous to the ecosystem, worldwide efforts have been made to find other means of protection against fish diseases, such as change in management procedures, genetic improvements, isolation as well as vaccination programs (Harell, 1979; Antipa, 1976; Fijan et al., 1977; Gunnels et al., 1976; Christensen, 1977; Hami and Kusuda, 1978).

Vaccines can be applied to fish orally added to the feed, by intraperitoneal injections, by dipping of the fish, or by the shower method. For mass vaccination the dip or shower method would probably be preferable because of the inconsistencies encountered with the oral method and the increased stress to the fish, and the high costs connected with the inoculation method (Harell, 1979). Hami and Kusuda (1978) claim that the spray method is less stressing to the fish than the dip method.

Vaccination is now the preferred method of vibriosis control among northwest-coast commercial salmon growers. According to Norwegian legislation of fish diseases, fish can only be vaccinated with vaccines approved by the Veterinary Division, Ministry of Agriculture. The purpose of the experiments described here was to compare and evaluate the effectiveness of a commercially available vibrio vaccine and a formalin-killed bacterin prepared at the National Veterinary Institute.

Materials and Methods

Facilities

The vaccines were tested in several fish farms. The immunization procedures were performed with the fish in freshwater either in fiberglass tanks or in earthen ponds.

Approximately 14 days post vaccination the fishes were transported to sea-water fish farms except in the cases when the vaccination was done at the weight stage of 1 g.

---
1 National Veterinary Institute, Oslo, Norway
2 Holmane Experimental Fish Farm, Sirevåg, Norway
3 Fish Breeding Experimental Station, Sunndalsøra, Norway

Fish

Rainbow trout *(Salmo gairdneri)* were used in the set-up, ranging in average weight from 1 to 300 g. These fishes were supplied from various freshwater fish farms during the period 1977–1979.

Vaccines

Bacterins were prepared at the National Veterinary Institute from *Vibrio anguillarum*, strain A 20/76, after 48 h growth at 24 °C on Trypticase soy broth (30 g per liter distilled water). The cells were killed with 0.5% concentrated formaldehyde for 1 h at 37 °C. The product contained approximately 10^{10} bacteria/ml. The commercially available vaccine Hivax from Tavolek Laboratories was used according to the recommendations of the manufacturer. It was applied either by dipping or by intraperitoneal (IP) injections, while the locally prepared bacterin was used only as IP, injections in doses of 0.5 ml per fish.

For marking and IP immunization Chlorbutol (300 mg/l) was used as anesthetic.

Recording of Mortalities

In some of the field trials quite exact recording of mortality was achieved throughout the period when the fishes were kept in sea water, while in the commercial fish farms only more approximate figures have been obtained.

The death rate in connection with vaccination procedures was negligible and the rainbow trout did not seem to suffer much from marking, handling stress, and injection.

Results

In one controlled field experiment which lasted for one year (1977–1978) the effect of Hivax and the produced bacterin was compared. Throughout the experimental period the fish suffered from spontaneous outbreaks of vibriosis. No medical treatment was given. Table 1 shows the mortality figures in each group of fish and the results clearly demonstrate a significant protection of the vaccinated fish against vibriosis.

In another series of controlled field experiments started in the spring of 1979 vaccination programs on different age groups of fish were compared. Table 2 shows no difference in mortality between fish vaccinated at the 100-g and at the 300-g stages, while fish vaccinated at the weight of 1 g showed the same mortality rate as the controls. The situation became very clear due to natural outbreaks of vibriosis at the end of July, approximately 1 month after transfer to sea water, and in mid-October (Figs. 1, 2). The control groups in cages 6 and 8 were left without antibiotic treatment while the fish in cage 9 (Hivax, 1 g) were medicated from 15–20 August and from 13–17 October. The medication was effective in reducing mortality rate in cage 9, but the mor-

Table 1. Mortality figures in a controlled field experiment 1977–1978

Experimental groups	Number of fishes	Average weight when vaccinated	No. dead fish from 31 Oct. 1977–30 Nov. 1978	Percent mortality
Controls	1881	75 g	472	25.1
Hivax IP	1812	75 g	228	12.6
Hivax dip	475	75 g	57	12.5
Norwegian IP	1832	75 g	178	9.7

Vaccination date: 31 October 1977
Transferred to sea water: 11 November 1977
Slaughtered: 30 November 1978

Table 2. Mortality figures in a controlled field experiment 1979

Experimental groups	Number of fishes in experiment	Average weight when vaccinated	Mortality figures from start to 22 Oct	Percent mortality
Hivax dip	900	100 g	7	0.78
Hivax IP cage 6	900	100 g	2	0.22
Norwegian IP	950	100 g	1	0.11
Controls	900	100 g	175	19.44
Hivax dip	700	300 g	2	0.29
Hivax IP cage 7	1200	300 g	5	0.42
Norwegian IP	1200	300 g	7	0.58
Controls cage 8	3100	300 g	370	11.94
Hivax dip cage 9	3366	Vaccinated 1978 at 1 g space weight, when transferred to sea water 130 g	409	12.15

Vaccination date: 20 May 1979
Transferred to sea water: 5 June 1979

The fish in cage 9 were treated with oxytetracycline from 15–20 August 1979 and from 13–17 October 1979. The other groups were left untreated

tality rate in the nonmedicated groups was also significantly lowered during the same period. Figure 3 shows the cumulative mortality within the various groups from the start of the experiment until now. The fish in cages 7 and 8 will be slaughtered in November 1979, while the fish in cages 6 and 9 will be kept in sea water until October – November next year, and this set-up might give some information on the lasting effect of vaccination throughout two summer seasons.

Table 3 shows the results of vaccination in several commercial fish farms. The figures show that vaccination seems to be an effective prophylactic treatment alone or in combination with medication.

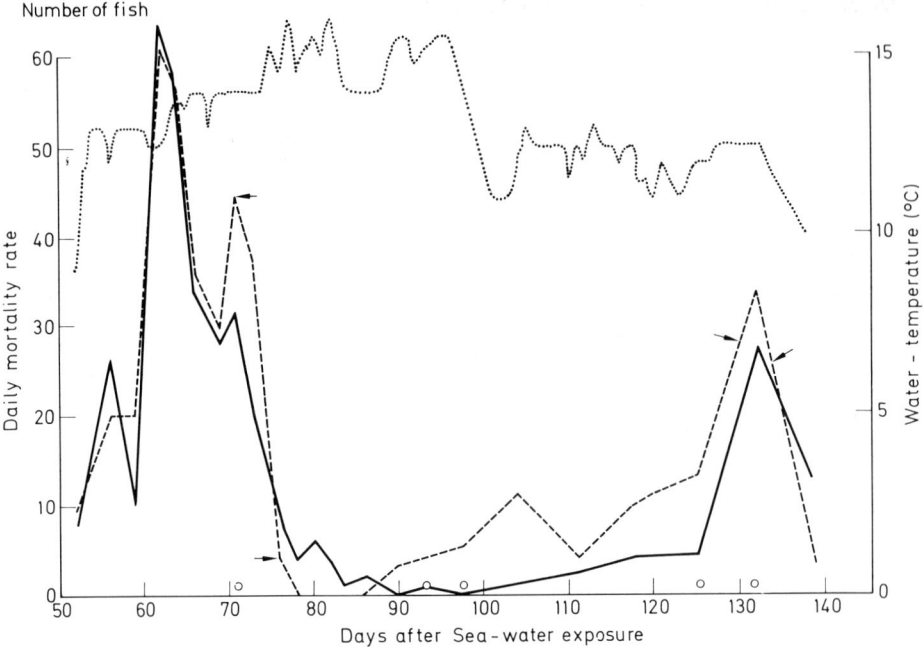

Fig. 1. Daily mortality rates of various groups of rainbow trout. Day 1 means day of transfer to sea water, ───── Control fish (300 g). ─ ─ ─ Fish dip-vaccinated in Hivax at the 1-g stage. These fishes were treated with oxytetracycline from 15–20 August and from 12–17 October. o Fish vaccinated at the 100- and 300-g stages with Hivax and our own bacterin. ... Sea-water temperature

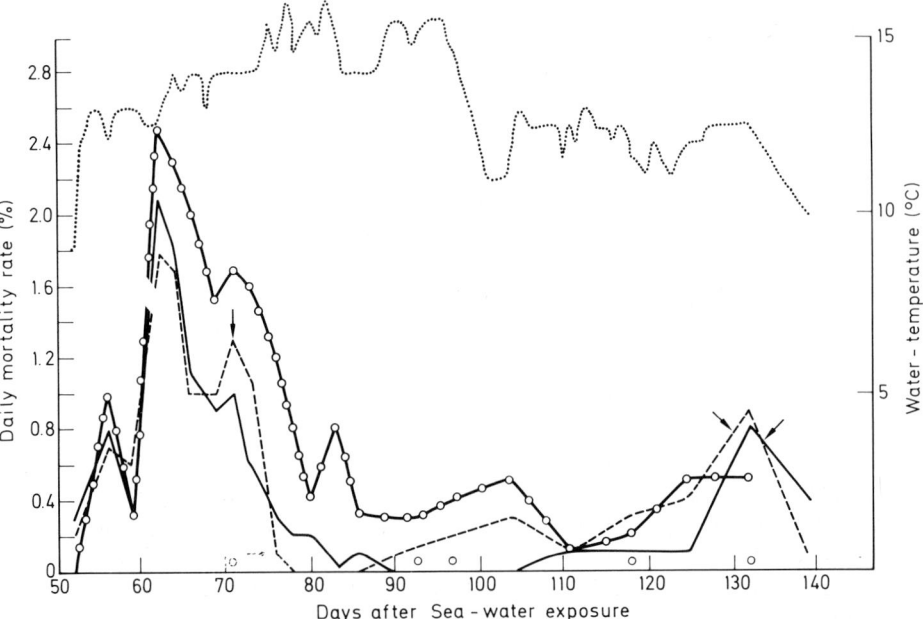

Fig. 2. Daily mortality rates in percent in various groups of rainbow trout after transfer to sea water. ───── Unvaccinated 300-g control group. o-o-o Unvaccinated 100-g control group. ─ ─ ─ Fish dip-vaccinated in Hivax at the 1-g stage. *Arrows* indicate start and end of antibiotic treatment. o Fish Vaccinated IP with Hivax or bacterin at the 100-or 300-g stages. ... Sea-water temperature

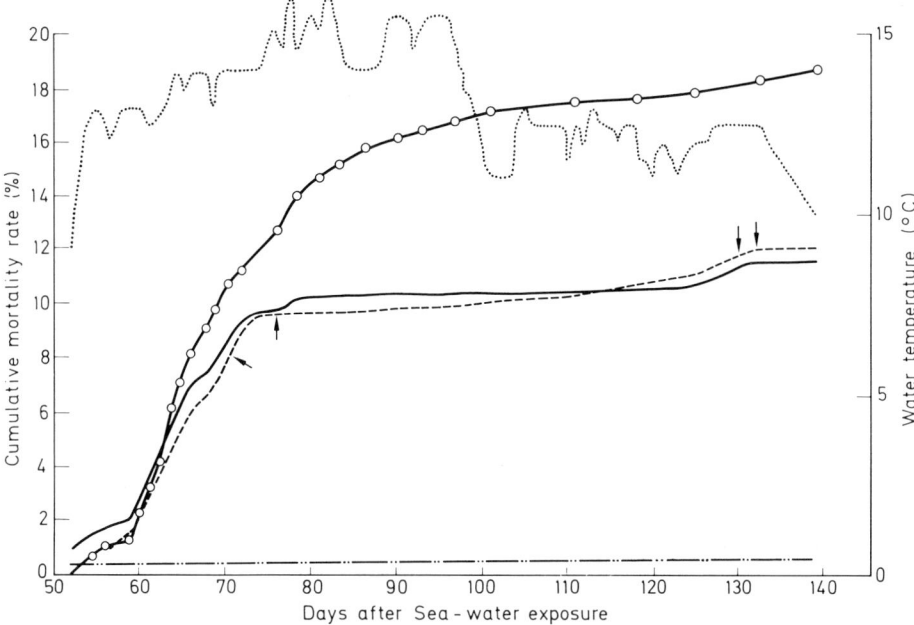

Fig. 3. Cumulative mortality rates in percent in various groups of rainbow trout after transfer to sea water. ──── Unvaccinated control group of 300 g. o-o-o Unvaccinated control group of 100 g. ––– Fish dip-vaccinated in Hivax at the 1-g stage. *Arrows* indicate start and end of antibiotic treatment. –··–··– Fish vaccinated with Hivax or bacterin at the 100- or 300-g stage by dipping or by IP injections. ... Sea-water temperature

Discussion

The survival rates are consistent with those experienced by other workers. Sawyer and Strout (1977) compared vaccination with medication in preventing vibriosis in coho salmon. Both methods caused increased survival rate and growth compared to the controls, and the total mortality rates were 2.9% in the vaccinated fish, 6.7% in the medicated ones, and 23.6% in the controls. Similar protection rates have been found in rainbow trout in vaccination experiments in Norway by Håstein et al. (1977), and by Egidius and Andersen (1979).

The last-mentioned authors used a *Vibrio anguillarum* bacterin prepared by trypsin digestion of the bacteria and a bathing period of 2 h. There was a mortality of 2.5% in the vaccinated group compared to 54% in the controls on challenge 4 months later. Similar mortality figures were recorded by Gould et al. (1979) when fishes vaccinated by the immersion method with a bivalent bacterin were challenged. In potency tests with Hivax, the challenge dose caused a mortality rate of more than 90% in the control groups and a corresponding 10% in the vaccinated group (Tavolek Technical Report, 1978).

Tavolek has also found similar mortality rates in fish dip-vaccinated at the 1-g stage and later challenged. The death rates recorded correspond well with our findings. Even if fish basically seem to be immunocompetent at the stage of approximately 1 g, vaccination at this stage does not give a long-lasting immunity. The mortality rate on

Table 3. Vaccination results of field trials in some commercial fish farms

Fish Farm	Number of fish Vaccinated	Number of fish Nonvaccinated	Average weight of fish when vaccinated	Vaccination period	Mortality rate	Approximate mortality %	Medical treatment
I	13,000	–	90 g	15 Oct 1977 –15 Sept. 1978	431	3.3	None
II	15,000	400–500	80 g	Autumn 1978–79	450	2.7	Yes. Due to "pollution" with small, nonvaccinated fish among the vaccinated fish
III	3,700	–	80 g	1977–78	75	2.03	None
IV	41,000	12 229	60 g	1977–78	40,000	Approximately 1% in vaccinated, 60% in nonvaccinated	None
V	Brood fish	–	–		None		None
VI	9,000	9,000	100 g 100 g	Spring 1979 Spring 1979	– 300	0.1 3.3	None Yes
VII	2,000 6,000		100 g 100 g	Spring 1979 Spring 1979			Yes None
VIII	100,000		1 g	1978	No records but vibriosis occurred		Yes
IX	15,000	20,000	100 g 100 g	Spring 1979 1979	–	0.33 2.5	None Yes

challenge seems to be the same as for unvaccinated rainbow trout (Downs, 1979). Since the cost of vaccination increases with increasing size of the fish, some experiments with vaccination of fish with Hivax at the 5-g stage have been started in 1979. These fishes have not yet been exposed to sea water.

However, vaccination of rainbow trout seems so far to give the best results when large fish are vaccinated some few weeks before they are transferred to sea water.

Though medication of rainbow trout against vibriosis offers as good results as vaccination, vaccination should be the future choice of treatment since the risks of development of drug-resistant pathogens and possible drug harm to consumers are avoided by this method.

What application method to be chosen for vaccination will depend on several factors such as number and size of the fish to be vaccinated and costs of vaccine.

Additionally, vaccination might be combined with genetic improvements as there is a significant difference in antibody titers in offspring after different fathers ($p<0.05$). (Refsti and Håstein, unpublished). It should therefore be possible to improve the resistance against vibriosis in rainbow trout by breeding. Better knowledge on the genetic parameters for this character is necessary before they can be included in the breeding programs.

References

Antipa R (1976) J Fish Res Board Can 33:1291–1296
Christensen N (1977) Riv Ital Piscic Ittiopat XII:23–25
Downs W (1979) Personal communication
Egidius EC, Andersen K (1979) J Fish Dis 2:405–410
Fijan N (1977) Pentrinoc Z, Stancl Z, Kezić N, Teskeredzić E Bull Off Int Epiz 87:439–440
Fryer JL, Rohovec JS, Tebbitt GL, McMichael JP, Pilcher KS (1976) Fish Pathol 10:155–164
Fryer JL, Amend DF, Harell LW, Novotny AJ, Plumb JA, Rohovec JS, Tebbit GL (1977) Oregon State University Sea Grant College Program. Publ No Oresu -T-17 012
Gould RW, Antipa R, Amend OF (1979) J Fish Res Board Can 36:222–225
Gunnels RD, Hodgins HO, Schiewe MH (1976) Am J Vet Res 37:737–740
Harell LW (1979) Personal communication
Hami T, Kusuda R (1978) Bull Jpn Soc Sci Fish 44:1413
Håstein T, Hallingstad F (1978) Unpublished
Håstein T, Refsti T, Gjerdrem T (1977) Bull Off Int Epiz 87:487–488
Refsti T, Håstein T (1979) Unpublished
Rohovec JS, Garrison RL, Fryer JL (1975) Proc 3rd US Jpn Meet Aquacult Spec Publ Jpn Fish Agencies Sea Regional Fish Res Lab, Niigato Jpn, pp 105–112
Sawyer ES, Strout RG (1977) Aquaculture 10:311–315
Sulimanovic D (1973) Vet Arch 43:153–161
Tavolek Technical Report (1978)
Vestergard-Jørgensen PE (1976) Nord Veterinaermed 28:570–571

Some Results of Vaccination Against Vibriosis in Brittany

F. BAUDIN LAURENCIN and J. TANGTRONGPIROS[1]

Although marine fish farming has been practised in France since the early 1970s, vibriosis was first reported in 1977. The infected animals were rainbow trout raised in brackish water. Since then, the number of reported cases in trout and other species has increased each year, although it still remains infrequent.

Preliminary experiments (Tangtrongpiros, 1978) demonstrated the efficiency in rainbow trout of a bacterin developed from the local strain. The vaccination was administered either orally or intraperitoneally and appeared effective in both cases.

This paper reports the results of a new laboratory experiment as well as of some field trials.

Materials and Methods

The Bacteria

The vibriosis isolated from diseased fish in Brittany are always similar, and thus designated *Vibrio anguillarum* 408, typical of the archetype proposed by Evelyn (1971). Presumptive identification is based on the following results: gram negativity, mobility and morphology under phase-contrast microscopy, oxidase activity, fermentation of glucose, sensitivity to the vibriostatic compound 0/129 (2,4 diamino-6,7 diisopropylpteridine).

A commercial multitube micromethod (API system for Enterobacteria) is used at 20 °C for practical, more accurate recognition. Using this method, we obtained the following results:

Nitrate reduction	+
ONPG test	+
Arginine hydrolysis	+
Lysine decarboxylation	−
Ornithine decarboxylation	−
Citrate assimilation	+
H_2S production	−
Urease activity	−
Tryptophan desamination	+
Indole production	+
Acetoin production	+

---

[1] Laboratoire National de Pathologie des Animaux Aquatiques, Ministère de l'Agriculture, Direction de la Qualité, Services Vétérinaires, COB-BP 337, 29273 Brest Cedex, France

Gelatin liquefaction	+
Acid from:	
Glucose	+
Mannitol	+
Inositol	–
Sorbitol	+
Rhamnose	–
Sucrose	+
Melibiose	–
Amygdalin	+
Arabinose	+

Using this technique, the bacteria appears identical to some Norwegian strains or slightly different from others (amygdalin and arabinose not acidified). In the same way, *V. anguillarum* 775 (Harrel et al., 1975) does not produce indole; a *V. anguillarum* isolated from rainbow trout in the United Kingdom (McCarthy et al., 1974) produced hydrogen sulfide.

The Vaccines

Three vaccines were used in the laboratory experiment:
V_1: heated bacterin developed in the laboratory from strain 408
V_2: formolized bacterin from the same strain, produced by a commercial company (IFFA-MERIEUX)
V_3: bacterin developed in the USA from strains 775 and 1669 (Harrel et al., 1975).
In the field trials, the V_2 vaccine was used alone.

Vaccination Procedures

Laboratory Experiment

At the time of vaccination water salinity was 10‰; this concentration was increased by 5‰ each week, to a maximum of 30‰. The temperature was approximately 12 °C throughout the experiment. A total of 48 tanks, each containing 25 coho salmon (mean weight 20 g), was subdivided into groups of 12 tanks. Each group corresponded to a different vaccine under study. The fourth group of tanks was reserved for the control animals.

    Within each experimental group of 12 tanks the following subdivision was made:
    4 tanks for oral administration: 2 mg of wet packed cells per day and per fish, for 12 days
    4 tanks for intraperitoneal injection: 3 mg of wet packed cells per fish in 0.1 ml
    4 tanks for administration by dipping: 2 min in hypersaline solution (53.2% NaCl), the 1 min in vaccinal bath (12 mg of wet packed cells per ml).

Field Trials

The vaccine used was in all cases the formalized bacterin produced by the commercial laboratory from strain 408.

1) June 1979, 2.500 rainbow trout (mean weight 200 g) kept in net cages in a Breton estuary (S:30‰; t°: 15 °C); oral distribution of 10 mg of wet packed cells per day per fish, for 15 days.
2) May 1979, 1.2 tons of coho salmon (mean weight 35 g), in freshwater; dipping method, as previously explained. The control fish were transported to the sea 2 weeks after the vaccination and the vaccinated fish 3 weeks after the vaccination.
3) June 1979, 80 rainbow trout (mean weight 250 g); dipping method. Half were placed immediately in sea water, half in freshwater, both with control fish.

Efficiency Tests

Agglutinin Titers

The blood was sampled via intracardial puncture using heparinized hematocrit tubes. Agglutinins were titered using microtiter plates: the agglutinin titer was arbitrarily defined as the inverse of the final dilution at which agglutination is observed.

Artificial Challenges

In the laboratory experiment, challenge was performed on all fish 3 months after vaccination, by IP injection in half of the tanks and per os administration in the other half. A 24-h culture of *Vibrio anguillarum* 408 used:
a) diluted in physiological saline (1.3×10^5 germs/ml) for intraperitoneal injection of 0.1 ml per fish
b) mixed in food pellets for a distribution of 6.7×10^{10} germs per fish per day during 5 days.

In the field trials, samples of vaccinated and control fish were taken to the laboratory and placed in separate tanks. Five control fish were then intraperitoneally inoculated and placed in each tank. Mortalities resulting from contamination were observed for 15 days. The diagnosis was made by isolation of the vibrio from the anterior kidney in trypticase soya agar 1.5% NaCl.

Results of Laboratory Experiment

Serum Antibodies

Comparison of Vaccines and of Vaccination Procedures

The results given in Table 1 and 2 are the means of the agglutinin titers from the six blood samplings taken after vaccination and before challenge; thus, each result involves about 48 fish. Following analyses of variance and multiple t-tests (significance level 0.05), nonstatistically different lots were grouped together. These lots showing statistical differences are indicated by capital letters A, B, C. ... It appears that:
a) There is little or no difference between the bacterins used (V_1, V_2, V_3)
b) Oral administration of antigen does not result in a change in the amount of agglutinins
c) Taking into account the quantities of bacterins used, dipping gives a lower level of agglutinins than that obtained via intraperitoneal injection.

Table 1. Antibodies: comparison between vaccines for each vaccination mode

Vaccination modes	Agglutinin titers according to vaccines			Control
	V_1	V_2	V_3	
Oral	A 37	A 44	A 42	A 34
IP Injection	B 82	A 186	A 160	C 28
Dipping	A 90	A 115	A 83	B 37

Table 2. Antibodies: comparison between vaccination modes, for each vaccine or for all the vaccines together

Vaccines	Agglutinin titers according to vaccination mode		
	IP injection	Dipping	Oral admin.
V_1	A 82	A 90	B 37
V_2	A 186	B 115	C 44
V_3	A 160	B 83	C 42
$V_1+V_2+V_3$	A 134	B 95	C 41

Table 3. Antibodies: chronological variations (the numbers preceded by the letter d indicate the number of days after vaccination

Control (all routes)	⇒	d 144 A 174	d 30 A 136	d 45 B 45	d 15 B 31	d 75 B 25	d 90 B 22	d 60 B 13
Oral administration $V_1+V_2+V_3$	⇒	d 30 A 192	d 144 A 142	d 45 B 43	d 75 B 31	d 60 B 30	d 15 B 29	d 90 B 21
IP injection $V_1+V_2+V_3$	⇒	d 30 A 292	d 75 A 261	d 144 A 192	d 60 A 161	d 45 A 160	d 90 B 56	d 15 B 54
Dipping $V_1+V_2+V_3$	⇒	d 30 A 330	d 45 B 154	d 60 C 88	d 144 C 88	d 75 C 61	d 90 C 57	d 15 C 60

Chronological Variations

The results are presented in Table 3, using the previously described statistical format. They represent each of the 6 blood samplings performed following vaccination (15, 30, 45, 60, 75, 90 days) and the last sampling performed months following challenge (day 144 after vaccination).

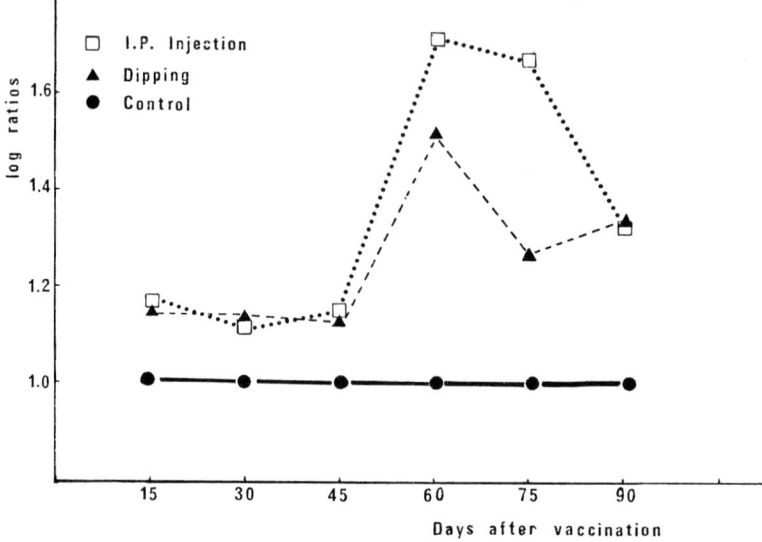

Fig. 1. Chronological variations of the log ratios of vaccinated fish titers to control fish titers

Considerable variations appear, even for control or orally vaccinated fish. There is an apparent heterogeneity of results with regard to chronology, and there is an essentially unexplained increase at day 30. However, it should be noted that the titers $1^1/_2$ months after challenge are very high, but higher for control or orally vaccinated fish; there is no appearance of a booster effect.

As a more meaningful method of presenting these results, we propose a chronological representation using the ratios of mean log agglutinin titers of inoculated or dipped fish to the mean log agglutinin titers of control fish (Fig. 1). Using this method, the difference in antibodies levels between control and vaccinated fish (inoculated or dipped) appears to increase after day 30 with a maximum at day 60, then to decrease until day 90.

Mortalities Following Challenges

The first challenge involved half of the different lots and was performed via intraperitoneal inoculation of virulent bacteria. The results are shown in Table 4.

The second challenge, concerning the other half of the lots, was an attempt at oral contamination (see Sect. 1.4.2, laboratory experiment). Mortalities are very low, both in vaccinated and control fish, as shown in Table 5. It is thus not possible to demonstrate a protector effect of the vaccines.

A protector effect was demonstrated, following intraperitoneal inoculation or dipping administration of the vaccines.

Table 4. Mortalities following the first challenge. Use of chi-squared test, significance level 0.001

Vaccination mode	Fish	Initial number of fish	Mortality (%)	Percent protection
Oral administration	Vaccinated	120	31.7	6.2
	Control	142	33.8	(not significant)
Intraperitoneal Inoculation	Vaccinated	140	1.4	96
	Control	142	33.8	(highly significant)
Dipping	Vaccinated	145	2.1	94
	Control	142	33.8	(highly significant)

Table 5. Mortalities following the second challenge

Vaccination mode	Fish	Initial number of fish	Mortalities after 2nd challenge
Oral administration	Vaccinated	140	4
	Control	142	4
Intraperitoneal inoculation	Vaccinated	118	0
	Control	142	4
Dipping	Vaccinated	142	0
	Control	142	4

Results of Field Trials

Oral Administration of Vaccine to Rainbow Trout

No increase of agglutinins was noted 2 months after vaccination. The mortalities after challenge were not statistically different between control (31.2%) and vaccinated fish (27.2%).

Dipping Administration of Vaccine to Young Coho Salmon

As shown in Table 6, there is a significant difference between control and vaccinated fish only until the 2nd month. It is interesting to note the very high titer 3 weeks after vaccination, just before the vaccinated fish were transported to the sea.

Table 6. Antibodies: chronological comparison between control and vaccinated fish (Field trial 2)

Fish	Time after vaccination			
	3 weeks	2 months	3 months	4 months
Control	B 21	B 17	A 47	A 34
Vaccinated	A 1973	A 82	A 49	A 60

A vibriosis appeared in the natural environment between the first and second blood samples. High mortalities appeared first in control fish, and two weeks later in vaccinated fish. However, both lots were treated with antibiotics, and mortalities were not significantly different.

However, the natural disease did not result in an increase of antibodies. Artificial challenges performed in the laboratory 4 months after vaccination produced some not easily interpretable results: higher mortalities in vaccinated fish (18.4%) than in control (5.2%) fish.

Dipping Administration of Vaccine to Rainbow Trout

Serum Antibodies

Agglutinin titers (Table 7) are significantly higher in vaccinated fish than in control fish after 2 months in the marine environment and after $3^1/_2$ months in freshwater.

Mortalities Following Challenge (Table 8)

The pathogenic effect was more pronounced in sea water than in freshwater. However, protection coefficients are highly significant in both environments.

Mortalities occurred from day 7 to day 14 after placing in tanks of 5 control fish inoculated with the virulent bacteria (i.e., day 5 to day 12 after the death of the first

Table 7. Antibodies: chronological comparison between control and vaccinated fish in marine and freshwater environments (Field trial 3)

Environment	Lots		Time after vaccination		
			1 month	2 months	3.5 months
Sea Water	Control	⇒	C 7.2	C 6.6	B 32
(S=30‰)	Vaccinated	⇒	A 105	B 26	B 35
Freshwater	Control	⇒	B 13	B 4	B 9
	Vaccinated	⇒	A 86	A 86	A 71

Table 8. Mortalities of rainbow trout (mean weight 300 g) following the "natural contamination" in sea water and freshwater (Field trial 3)

Environment	Fish	Initial Numbers of Fish	Mortalities (%)	Protection
Sea water	Control	29	86.2	52
	Vaccinated	29	41.4	
Freshwater	Control	30	23.3	100
	Vaccinated	35	0	

inoculated fish). During the challenge, the temperature fluctuated between 12° and 14°C.

Conclusions

The tested vaccines showed a comparable immunitary effect, whatever the strain or the method of production.

The oral mode of administration, as is generally recognized, did not increase the level of agglutinin. However, no protection could be established, contrary to the results obtained by various authors, notably the previous observation of Tangtrongpiros (1978). This result could be due to the mode of challenge in the laboratory experiment (IP inoculation) but not to the "natural contamination" used in field trial 1.

The other methods of vaccination produced an increase in the agglutinin titers, but these were usually not very high, and in any case often varied from fish to another. The highest titers were obtained after 2 months in the laboratory experiment (about 300) and in field trial (about 100). In field trial 2 (dipping of coho salmon) an agglutinin titer of about 2,000 was obtained 3 weeks after vaccination; nevertheless, considerable mortalities occurred soon after, due to the appearance of a natural vibriosis. In this case, a considerable pathogenicity of *Vibrio* or the stress effect of the transport from freshwater tanks to the sea water could be responsible. However, the large decrease of agglutinin titer after the disease should be noted. It is also interesting to note that the challenge did not induce a higher agglutinin level in the vaccinated fish than in the control (i.e., no booster effect). With respect to environment, field trial 3 showed a briefer elevation of agglutinins in freshwater than in sea water.

The challenges were usually carried out when the agglutinin levels had decreased. However, the protection remained satisfactory, although it is difficult to compare one experiment to another because of the differences in pathogenic effects. In the laboratory experiment it appears that the dipping mode of vaccination provided an obviously similar protection to that provided by the intraperitoneal inoculation. In addition, the field trials demonstrated the feasibility of the dipping mode both for small fish (35-g coho salmon) and for larger fish (250-g rainbow trout).

References

Anderson DP (1974) Fish immunology, T.P.H. Publication Inc Neptune, New Jersey USA
Antipa R (1976) Testing of injected *Vibrio anguillarum* bacterins in Pen-Reared Pacific salmon. J Fish Res Board Can 33:1291–1296
Antipa R (1976) Testing of injected *Vibrio anguillarum* bacterins in Pen-Reared Pacific salmon. jection and hyperosmotic infiltration of *Vibrio anguillarum* and *Aeromonas salmonicida*. J Fish Res Board Can 34:203–208
Baudin-Laurencin F (1977) L'élevage du saumon coho en France: Problèmes pathologiques. Océanis 3:267–270
Egidius E, Anderson K (1977) Norwegian reference strains of *Vibrio anguillarum*. Aquaculture 10:215–219
Evelyn TPT (1971) First records of Vibriosis in Pacific salmon cultured in Canada and taxonomic status of the responsible bacterium, *Vibrio anguillarum*. J Fish Res Board Can 28:517–525
Evelyn TPT (1977) Immunization of salmonids. Proceedings of the international symposium on diseases of cultured salmonids. Tavolek Inc Seattle USA, pp 161–176
Fryer JL, Rohovec JS, Tebbit GL, McMichael JS, Pilcher KS (1976) Vaccination for control of infectious diseases in Pacific Salmon. Fish Pathol 10:155–164
Glorioso JC, Amborski RL, Larkin JM, Amborski GE, Celley DC (1974) Laboratory identification on bacterial pathogens of aquatic animals. Am J Vet Res 35:447–450
Harrel LW, Novotny AJ, Schiewe MH, Hodgins HO (1975) Two different Vibrios implicated in extensive mortalities of Pacific salmon cultured in marine waters of Puget Sound, Washington. National Oceanic and Atmospheric Administration, 2725 Montlake Boulevard East, Seattle, Washington 98112
McCarthy DH, Stevenson JP, Roberts MS (1974) Vibriosis in rainbow trout. J Wildl Dis 10:2–7
Michel C (1979) Furunculosis of salmonids: Vaccination attempts in rainbow trout *(Salmo gairdneri)* by formalin-killed germes. Ann Rech Vet 10:33–40
Tangtrongpiros J (1978) Essai de vaccination contre la vibriose à *Vibrio anguillarum* 408 chez la truite arc-en-ciel *(Salmo gairdneri)*. Mémoire du DEA Faculté des Sciences Université de Bretagne Occidentale

Production and Properties of a Haemolytic Toxin by Vibrio anguillarum

C. B. MUNN[1]

Introduction

Vibrio anguillarum is an important pathogen of fish, especially marine fish reared under intensive conditions [2, 6]. The economic importance of this disease has prompted considerable research into the development of vaccines for its control, and these have met with partial success [3, 7]. However, very little is known of the biochemical basis of pathogenicity of *V. anguillarum*. As well as the scientific value of understanding the fundamental mechanisms underlying the interactions between pathogens and their hosts [12], such knowledge can have practical importance, especially in the development of improved and more specific vaccines.

In those bacterial diseases of man and animals which have been well investigated, the production of a protein toxin or toxins by the pathogen has often been shown to play a major role in pathogenesis [1].

The anaemic response of fish led Roberts [11] and Wolke [16] to postulate that a haemolytic toxin played a key role in the pathogenesis of vibriosis. Most strains of *V. anguillarum* are haemolytic when grown on blood agar, and the haemolysins and other membrane-damaging toxins of many other bacteria have been isolated and proved to be toxic [4, 9]. However, I would like to stress the difficulty of relating effects observed in vitro with those in vivo [13], and considerable effort is required to prove a causal role for a toxin in the production of disease symptoms during natural infection.

In a previous paper [10] I described the isolation and partial purification of *V. anguillarum* haemolysin. In this paper, I describe preliminary results of further studies on this protein toxin.

Materials and Methods

Isolation and Partial Purification of Haemolysin

V. anguillarum was cultivated in shake-culture for 22 h in a peptone-yeast extract medium, and crude haemolysin was prepared as previously described [10]. Gel chromatography was carried out on Sephacryl S-200 (Pharmacia) in sodium acetate (0.1 M), sodium chloride (0.2 M), pH 5.5 in a 90x1.5 cm column.

---

1 Plymouth Polytechnic, Plymouth PL4 8AA, England

Haemolysin Assays

Erythrocytes were washed thoroughly in phosphate-buffered saline (PBS) and resuspended to a density which gave an absorbance at 540 nm of 0.8 when lysed with an equal volume of SDS (0.025%). Assays were performed in microtitre trays, 0.1 ml of erythrocyte suspension being added to 0.1 ml of diluted haemolysin, incubated at 37 °C for 1 h, and the 50% endpoint estimated visually.

Kinetics of Lysis

For kinetic studies of lysis, equal volumes (1.8 ml) of haemolysin and standardized erythrocyte suspension were mixed in 15-ml conical centrifuge tubes. Tubes were removed from incubation at various times and quickly centrifuged (1 min, 3000 r.p.m.). The absorbance of the supernatant was measured at 540 nm. Percentage haemolysis was calculated relative to cells totally lysed with an equal volume of SDS.

In Vivo Effects

Dilutions of toxin (0.1 ml) were injected into the dorsal aorta of groups of 3 unanaesthetized eels (approximately 60 g) in 50% sea water. Control fish were injected with 0.1 ml PBS.

Results and Discussion

Isolation and Purification

A major problem in the isolation and purification of *V. anguillarum* haemolysin has been its extreme lability. Even under cold-room conditions at a favourable pH [10] losses during purification were considerable. It was thought that this might be due to degradation by contaminating proteolytic enzymes. However, incubation with chelating agents (EDTA, nitrilotriacetate) and the protease inhibitor phenylmethyl sulphonyl fluoride did not reduce the loss. Partial purification by ammonium sulphate precipitation followed by gel chromatography has been achieved, but it has not yet been possible to purify the toxin further because of the low yield. Figure 1 shows a typical separation on Sephacryl S-200. Comparison of the elution characteristics of this peak with standard marker proteins gave a molecular weight (very approximate) of 191,000. In some separations the peak showed considerable "tailing" reflecting degradation to lower molecular weight components. Although the purity of this material is satisfactory for many purposes, it is now recognised that for detailed study of the biochemical properties of toxins, very extensively purified proteins should be used [15]. Therefore, some caution is needed in the interpretation of the present results.

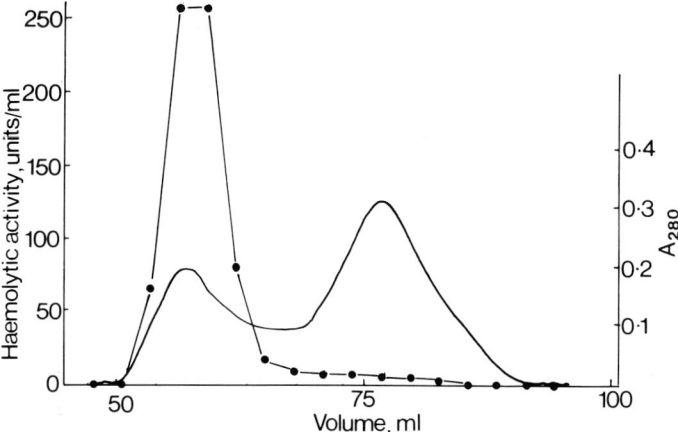

Fig. 1. Separation of haemolysin from crude ammonium sulphate preparation on Sephacryl S-200. —— absorbance, •——• haemolytic activity against horse erythrocytes

Toxic Effects on Fish

Eels injected with the highest concentration of haemolysin used (ca. 1,000 units/ml) died within 20 min of injection. Before death, the fish showed violent spasmic contractions of the body. Death was sudden and characterised by complete flaccidity of the body muscle. At lower doses, death occurred within 2–4 h. The symptoms observed suggest that this toxin preparation may possibly have neurotoxic activity.

Kinetics of Lysis

In order to determine whether the haemolysin behaves as a typical bacterial cytolytic toxin, the kinetics of haemolysis were followed. Horse erythrocytes were used for these experiments. The proportion of cells lysed at various times of incubation was followed by spectrophotometry of the haemoglobin released. Results show that haemolysis is a two-stage process consisting of an initial pre-lytic phase followed by a phase during which actual cell lysis and release of haemoglobin occurs. Both the length of the pre-lytic phase, and the rate of haemoglobin release are dependent on haemolysin concentration (Fig. 2) and on temperature (Fig. 3). No haemolysis was detected within 4 h at temperatures below 10 °C. The pre-lytic phase also does not seem to occur at low temperatures, because erythrocytes incubated with haemolysin at 0 °C for 4 h before transferring to 37 °C still required the same pre-lytic period before haemoglobin release commenced. The results are rather unusual compared with other cytolytic toxins and these effects should be reinvestigated with haemolysin of higher purity. The high optimum temperature for haemolysis is perhaps surprising in view of the optimum temperature of 20 °–22 °C for growth of *V. anguillarum*. The nature of the pre-lytic phase was investigated by treating erythrocytes with haemolysin for various times, centrifuging quickly, resuspending in diluent and continuing incubation, so that cells

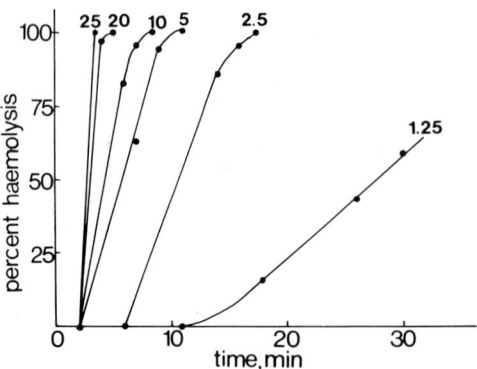

Fig. 2. Effect of haemolysin concentration (units/ml) on rate of haemolysis of horse erythrocytes

were removed from contact with free haemolysin after various periods. Even after only 5 min exposure to the haemolysin, most of the cells ultimately lysed. In another experiment, the supernatant fluid from cells treated with haemolysin for 10 min was added to fresh, untreated erythrocytes. The haemolytic activity of the supernatant fluid was considerably reduced, indicating that much of the haemolysin is either inactivated or bound irreversibly during the early part of the pre-lytic phase. The kinetics of lysis appear to be somewhat unusual and further analysis is required.

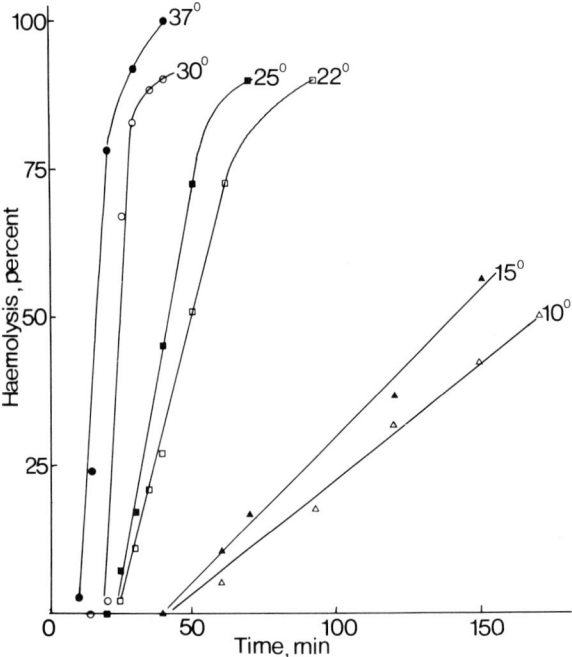

Fig. 3. Effect of temperature (°C) on rate of haemolysis of horse erythrocytes

Inactivation of Haemolysin by Gangliosides

The related species *V. cholerae* [8] and *V. parahaemolyticus* [14] both produce toxins which have been shown to bind to the membrane glycolipids known as gangliosides. It was therefore decided to determine whether *V. anguillarum* behaved similarly. Various concentrations of haemolysin and bovine ganglioside mixture were mixed in microtitre trays in a chessboard titration. After incubation (37 °C, 10 min) horse erythrocytes were added and incubation continued for 1 h. Haemolysin was inactivated by gangliosides, and Fig. 4 shows that there is a linear relationship between the activity of haemolysin and the concentration of gangliosides required to inactivate it. Incubation of erythrocytes with gangliosides for 3 h followed by thorough washing before addition of haemolysin resulted in marked reduction in their susceptibility to lysis. This result is rather surprising but might be explained if exogenous ganglioside becomes associated with the membrane and competes with the natural binding sites. Since only a crude mixture of gangliosides was used, it is not possible to identify at present the particular glycolipid structure with which haemolysin interacts. However, it does not appear to be one of the neuraminidase-sensitive gangliosides, since pre-incubation of ganglioside mixture with this enzyme did not reduce the ability of the ganglioside to inactivate haemolysin. In order to investigate the possibility that the ganglioside G_{M1} is the specific receptor, ganglioside mixture was incubated with excess cholera toxin before adding to haemolysin. The inactivating ability of ganglioside was not reduced by cholera toxin. Since cholera binds specifically and precipitates ganglioside G_{M1} [8] it seems unlikely that this is the receptor for *V. anguillarum* haemolysin. Future work will be

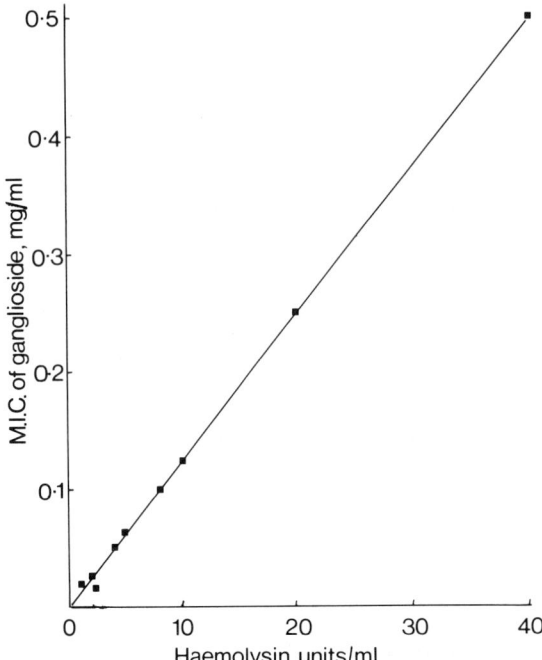

Fig. 4. Inactivation of haemolysin by ganglioside mixture

aimed at further elucidation of the mode of action of the haemolysin and the nature of possible receptors.

Role in Pathogenicity

It is not yet possible to determine what role the haemolysin plays in pathogenicity. We have been unable to demonstrate a link between virulence and production of the haemolysin in vitro. Also, haemolysin production is not coded by plasmid genes (Munn and Pearcey, in preparation) whereas Crosa et al. [5] claimed that virulence was plasmid-coded. However, there is some evidence that haemolysis does occur in vivo, the haemolysin is lethal to eels, and some of the symptoms observed here are similar to those seen in infected eels (H. Chart, personal communication). However, the complexity of host-pathogen relationships is increasingly recognised and this toxin is probably only one of several factors which determine pathogenicity.

Acknowledgements. I thank Robert Pearcey for technical assistance and Henrik Chart for performing the in vivo experiments. The work was supported by the Science Research Council.

References

1. Ajl SJ, Kadis S, Montie TC (1970) Microbial toxins, vols I, 2 A. Academic Press, London New York
2. Anderson JW, Conroy DA (1970) Vibrio disease in fishes. In: Snieszko SF (ed) Diseases of fishes and shellfishes. American Fisheries Society, Washington, pp 266–272
3. Antipa R (1976) Field testing of injected *Vibrio anguillarum* bacterins in pen-reared Pacific salmon. J Fish Res Board Can 33:1291–1296
4. Bernheimer AW (1970) Cytolytic toxins of bacteria. In: Ajl SJ, Kadis S, Montie TC (eds) Microbial toxins, vol I. Academic Press, London New York, pp 183–209
5. Crosa JH, Schiewe MH, Falkow S (1977) Evidence for plasmid contribution to the virulence of the fish pathogen *Vibrio anguillarum*. Infect Immun 18:509–513
6. Fryer JL, Nelson JS, Garrison RL (1972) Vibriosis in fish. In: Moore RW (ed) Prog Fish Food Sci 5:129–133
7. Fryer JL, Rohovec JS, Garrison RL (1978) Immunization of salmonids for control of vibriosis. Mar Fish Rev 40:20–23
8. Holmgren J (1978) Cholera toxin and the cell membrane. In: Jeljaszewicz J, Wadström T (eds) Bacterial toxins and cell membranes. Academic Press, London New York, pp 333–366
9. Jeljaszewicz J, Szmigielski S, Hryniewicz W (1978) Biological effects of staphylococcal and streptococcal toxins. In: Jeljaszewicz J, Wadström T (eds) Bacterial toxins and cell membrans. Academic Press, London New York, pp 185–219
10. Munn CB (1978) Haemolysin production by *Vibrio anguillarum*. FEMS Microbiol L 3:265–268
11. Roberts RJ (1976) Bacterial diseases of farmed fishes. In: Skinner FA, Carr JG (eds) Microbiology in agriculture fisheries and food. Academic Press, London New York, pp 55–61
12. Smith H (1968) The biochemical challenge of microbial pathogenicity. Bacteriol Rev 32:164–184
13. Smith H, Taylor J (1964) Microbial behaviour in vivo and in vitro, CUP, Cambridge
14. Takeda Y, Takeda T, Honda T, Miwatani T (1976) Inactivation of the biological activities of the thermostable direct haemolysin of *Vibrio parahaemolyticus* by ganglioside G_{T1}. Infect Immun 14:1–5
15. Wadström T (1978) Advances in the purification of some bacterial protein toxins. In: Jeljaszewicz J, Wadström T (eds) Bacterial toxins and cell membranes. Academic Press, London New York, pp 9–49
16. Wolke RE (1975) Pathology of bacterial and fungal disease affecting fish. In: Ribelin WE, Magaki G (eds) Pathology of fishes. Univ Wisconsin Press, pp 33–117

Observations on Vibriosis in Cultured Flatfish

R. H. RICHARDS[1]

Introduction

The farming in Scotland of marine flatfish, in particular the turbot *Scophthalmus maximus* and the Dover sole *Solea solea* has been taking place since the 1960s. The principal restraints to future development are the supply of young fish, development of adequate diets and suitable husbandry techniques and the control of disease.

Epizootics of vibriosis have occurred amongst a number of species of wild fish (see e.g. Bagge and Bagge, 1956; Kusuda, 1966; Cisar and Fryer, 1969) and vibriosis ranks as a major cause of mortality amongst farmed flatfish (Anderson and Conroy, 1968, 1970; Horne et al., 1977) and in the marine culture of salmonids (Rucker, 1959; Wood, 1974, Cisar and Fryer, 1969).

The present paper describes the clinical appearance and histopathology of the condition in cultured turbot and Dover sole.

Materials and Methods

Samples of turbot and Dover sole, varying from 0+ to 3+ in age and from 2 g to 2,000 g in weight were routinely examined between April 1976 and October 1979 as part of a disease control service at at a site utilising the heated effluent from a power station on the West coast of Scotland. Fish were ongrown at temperatures ranging between 18 °C in summer and 13 °C in winter.

Histological samples of all organs were processed routinely, embedded in paraffin wax and sections cut at 5 μ. Stains employed included H and E, Gram-Humberstone and Perl's Prussian Blue method for the demonstration of ferric iron.

Results

Gross Lesions

First frequently died at all stages of infection but the presence of intercurrent disease, such as *Trichodina*, *Herpesvirus scophthalmi* or Haemogregarina infection, served to increase incidence and severity of signs.

---

1 Institute of Aquaculture, University of Stirling, Stirling FK9 4LA, United Kingdom

Gross lesions and clinical signs also varied with the species of fish, age and temperature, but could be broadly classified as follows:

1. Peracute. Death often occurred with no visible gross changes except for a darkening of colouration. In young 0+ turbot, peri-orbital oedema and gut oedema with copious quantites of peritoneal fluid were also found. This resulted in gross abdominal distension and typical white circlets of tissue surrounding the eye. Skin lesions were absent. Following a peracute outbreak, an acute syndrome often developed.

2. Acute. Two main classes of lesion were generally found. In both cases fish were inappetant, dark in colour and lethargic. Erosions of the jaw area were frequently found as a primary lesion in turbot of all age-classes. *Cytophaga* spp. organisms were usually present in such areas in large numbers, as well as a range of other bacteria including a variety of *Vibrio* spp. The large numbers of *Cytophaga* often led to a distinct yellow or orange colouration of the lesion. Further lesion development then principally involved either skin and superficial musculature, or internal organs.

Skin lesions usually commenced with an increased quantity of mucus and distinct haemorrhages in a number of areas, but principally involving the fins. In more long-standing cases, sloughing of areas of skin associated with haemorrhage was often seen. Gross internal lesions were seldom seen and lesions progressed more rapidly at higher temperatures. Fish were also affected with *Vibrio* spp. infections without marked skin lesions. In these cases haemorrhages were evident in many of the internal organs, the gut was often inflamed and contained a thick mucus, and quantities of straw-coloured or blood-stained ascitic fluid were found. The gills were usually pale in colour.

3. Chronic. Chronic skin lesions were generally deep ulcers with marked haemorrhage and fibrin deposition. On occasion, ulceration penetrated through the abdominal wall, leading to protrusion of the viscera through the fistula. In Dover sole, areas of dark-coloured necrotic skin developed very rapidly, forming the condition now commonly referred to as "black patch" disease. In systemic cases, peritoneal fluid still remained and fish often showed peritoneal adhesions. Gills were again pale in colour and chronic infections of the eye often led to evulsion of orbital contents.

Histopathology

1. Peracute. In juvenile turbot, there was a pronounced submucosal oedema of the gut with congestion and occasional haemorrhage, but the epithelium appeared normal. Muscular layers were variably affected with oedematous change (Fig. 1 a). Oedema of the choroid and the muscular and fibrous tissues surrounding the orbit was also present with spongiotic changes in overlying epidermis. This process extended around nerve trunks where it was often associated with an inflammatory response. Retinal oedema led to retinal detachment at the junction between the neural layers and the pigment epithelium.

The atrium of the heart showed myofibrillar necrosis with sloughing of the atrial endothelium. Liver showed areas of focal necrosis (Fig. 1 b) and varying degrees of fatty degeneration, but the latter change was also variably present in apparently healthy fish and thought to be associated with dietary imbalance. Haematopoietic tissue was

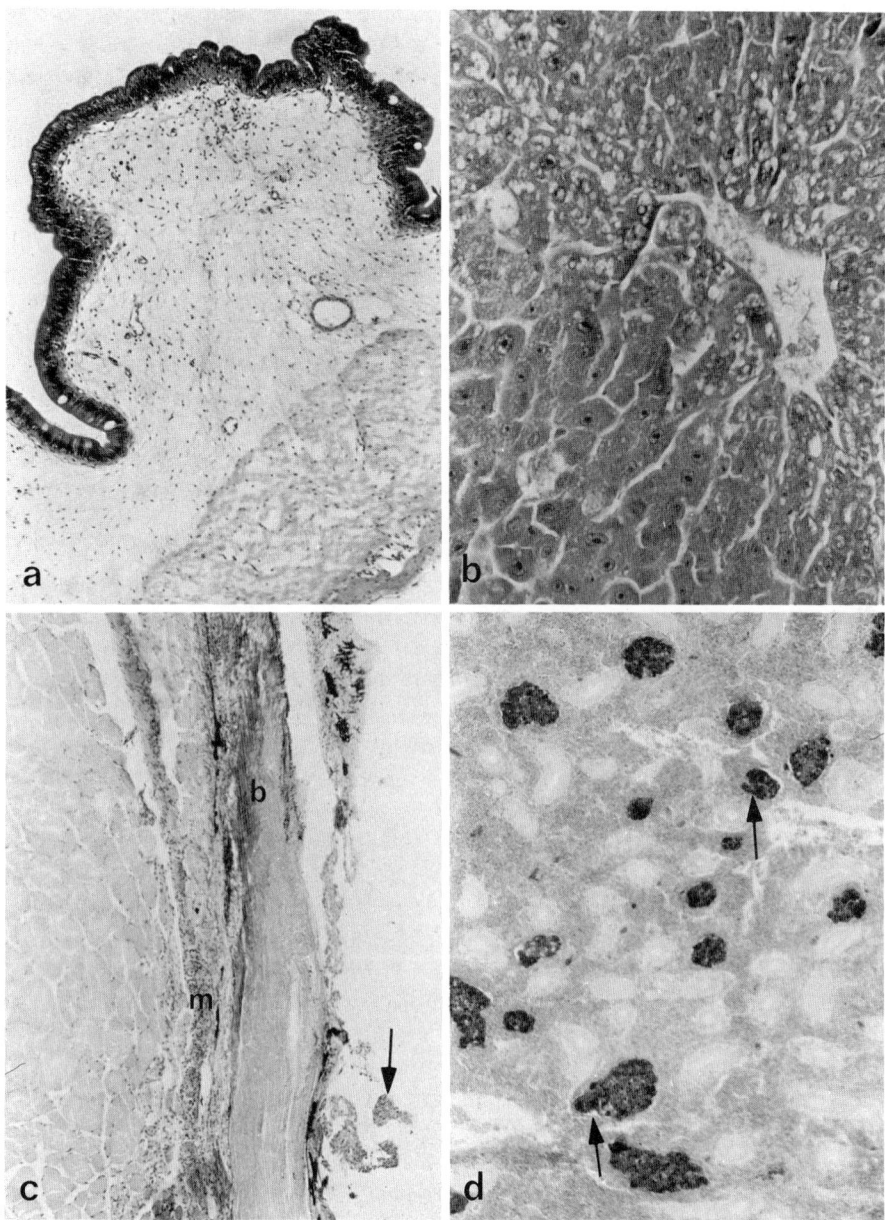

Fig. 1. a Gut oedema. H and E x125. Note submucosal and muscle oedema. b Liver. H and E x320. Extensive focal necrosis. c Integument. H and E x50. Note sloughing epidermis *(arrowed)*, extensive hypodermal bacterial infection *(b)* and muscular inflammatory response *(m)*. d Kidney. Perl's stain x125. Note iron deposits in melanomacrophage centres *(arrowed)*

reduced and often necrotic in both spleen and kidney, and destruction of melanomacrophage areas was observed. Vacuolation and necrotic change were found in renal tubular epithelial cells but there was little associated inflammatory response. Meningeal oedema occasionally occurred.

2. Acute. In those fish with skin and fin lesions, epidermis became spongiotic, necrotic and eventually sloughed and there was associated hyperaemia, oedema and haemorrhage in the dermis and hypodermis. Large numbers of bacteria were usually present in the hypodermis and on the surface of lesions (Fig. 1 c). In deeper ulcers, areas of necrotic muscle with an associated inflammatory response of mixed cell-type, prominent myophagia and variable numbers of bacteria were present.

The kidney and spleen showed changes similar to those in the peracute stage together with increased iron deposits in melanomacrophage centres (Fig. 1 d). Heart changes were similar to the peracute condition. Inflammatory responses were much more marked at higher temperatures.

3. Chronic. Skin lesions had often progressed to extensive areas of ulceration and muscle necrosis was advanced. Inflammatory infiltrates were pronounced with greater numbers of lymphocytes present at this stage.

Greatly increased quantities of iron were present in melanomacrophage centres of spleen, but atrial lesions had regressed. Considerable haematopoietic proliferation was evident in spleen and kidney, and in some cases, haematopoiesis was also present in liver or epicardium.

Attempts at healing with epidermal proliferation were evident at the edges of the ulcers but this did not cover the defect when associated with heavy bacterial infection.

Prophylaxis and Treatment

The principal area of consideration in prophylaxis is the reduction of stress and maintenance of good hygiene. Minimum handling of fish and regular cleaning of holding facilities led to a reduction in disease incidence. Provision of hatchery stock rather than wild caught fish has helped to reduce traumatic skin lesions which otherwise rapidly resulted in bacterial contamination and death.

A number of antibiotic treatments have been successfully employed, but their success depended on the extent of skin ulceration. With dietary treatment success depended on the avidity with which fish fed.

Oxytetracyclines have been incorporated into the diet at a dose rate of 80 mg/kg fish/day for 10 days but were obviously of little use if fish were not feeding. Bath treatment was used successfully with young fish at a concentration of 25 ppm for 30 min. Nitrofuran compounds in the food, e.g. Neftin at 100 mg active ingredient/kg fish/day for 10 days, have been similarly successful, and furanace baths at 1 ppm for $1/2$ h. have also proved beneficial. Policy has always been to employ a full course of treatment and use one type of antibiotic at a time to reduce the development of antibiotic resistance.

In the case of "black patch" disease in Dover sole, provision of sand in which fish can burrow on the tank bottom has dramatically decreased in incidence of the disease.

Reduction of temperature was occasionally beneficial but care must be taken not to reduce feeding response.

Discussion

The aetiology of vibriosis in marine flatfish is complex. Outbreaks often occur following netting of wild 0+ stock and subsequent transfer to ongrowing facilities. Skin lesions caused during netting allow entry of a variety of pathogenic organisms and strains of *V. anguillarum* and other *Vibrio* species readily proliferate in such areas. Particularly pathogenic strains of *V. anguillarum* have always caused systemic vibriosis and strains with different temperature optima have been isolated. It has recently been found that initial isolation of some pathogenic strains of *Vibrio anguillarum* depends upon the presence of blood in the isolation medium (Horne, personal communication 1979). The strains involved are probably opportunist invaders, the type of bacteria present depending on the skin (or gut) flora, which is generally a reflection of the bacteriological status of the environment (Evelyn and McDermott, 1961; Horsley, 1973). Transfer of fish from one locality to another may expose them to new strains of bacteria to which they have little immunity, although there is evidence that amongst the genus *Vibrio* there may be homologous plasmids responsible for pathogenicity (Crosa et al., 1977). As fish are fed on marine offals, there is also the likelihood of introducing quantities of a variety of *Vibrio* spp. via the food, suspected as a cause of *Vibriosis* in freshwater farming systems.

A number of species and strains of *Vibrio* are apparently involved in causing disease in flatfish. It has been suggested that the marine vibrios form a spectrum of organisms with heterogeneous properties rather than a number of well-defined species (Baross et al., 1978) but the isolation of types from diseased turbot serologically identical with strain 1669 affecting salmonids in the Pacific Northwest (Novotny, personal communication) demonstrates a widespread occurrence and an apparent lack of host-specificity, although for instance the haemolysins produced may show a variation in effect between species (McArdle, 1973).

At the farm in question, heated effluent from a power station is mixed with ambient sea water to produce higher mean temperatures suitable for increased growth. Unfortunately, pump failures occasionally lead to rapid temperature fluctuations, and under such conditions outbreaks of vibriosis rapidly develop. Chlorination is used to control pollution in piping systems. At low levels this is advantageous in controlling populations of bacteria and other pathogens. On occasion, however, flushes of chlorine of higher concentration pass through the farm and these have led to gill and skin damage resulting in outbreaks of disease.

Traumatic damage occurs during netting and the nature of the tank surface may also be responsible for early damage to the jaw area which often precedes a clinical outbreak. The reduction in the incidence of "black patch" disease by the provision of sand may be due to the removal of excess mucus and bacteria from the skin surface. Healing certainly improves considerably when excessive mucus and dead tissue are removed, as in the treatment of salmonid bacterial gill disease (Cawley, 1976).

Vibriosis in marine flatfish has been described on a number of occasions but with the exception of reports from Britain (Anderson and Conroy, 1968; Horne et al., 1977) most reports have been concerned with wild fish (e.g. Levin et al., 1972). The peracute syndrome in juvenile turbot was essentially as previously described by Horne et al. in 1977 with the exception of marked gut oedema seen since that date. Such a syndrome has so far only been noted in juvenile turbot. The oedema is thought to result from cardiac failure and perhaps an effect on capillary permeability rather than through renal failure (Horne et al., 1977). Except for the work of Horne et al. (1977) only an acute to chronic condition involving skin ulceration with the development of anaemia has been described. Although no haematological measurements were made in the present work, evidence of damage to haematopoietic tissue and accumulation of iron deposits predominantly in the spleen [similar to those described by Agius (1979)] suggests anaemia with compensatory haematopoiesis in the more chronic condition. Skin lesions were similar to those described by Hodgkiss and Shewan (1950), Anderson and Conroy (1970), Levin et al. (1972), and McArdle (1973). Inflammatory response described by Levin et al. (1972) consisted principally of lymphocyte infiltration. The responce in the present work was more pronounced at higher temperature and was of mixed cell-type.

Healing of skin wounds would appear to depend on the absence of necrotic material and bacteria and the effects of temperature. Many vibrio strains are more active at higher temperatures (i.e. 15°–20 °C) and a reduction in temperature, though reducing the speed of healing, markedly inhibits proliferation of such strains. It is for this reason that a temperature reduction often aids recovery, though strains active at lower temperature have been found, parallelling findings in the USA (Harrell et al., 1976; Sawyer, personal communication 1976).

Death in vibrio outbreaks is thought to result from toxic damage to parenchymal organs (Umbreit and Tripp, 1975) or through anaemia (Tanaka, 1974). It also seems extremely likely that death from osmotic imbalance such as occurs in *Saprolegnia* infections in salmonids (Richards and Pickering, 1979) is a common sequel to extensive ulceration and loss of skin surface.

Vibriosis may be controlled by careful husbandry practice, prompt and rational use of antibiotics and perhaps, in future, through vaccination programmes.

References

Agius C (1979) The role of melano-macrophage centres in iron storage in normal and diseased fish. J Fish Dis 2:337–343

Anderson JIW, Conroy DA (1968) The significance of disease in preliminary attempts to raise flatfish and salmonids in sea water. Bull Off Int Epiz 69(7–8):1129–1137

Anderson JIW, Conroy DA (1970) Vibrio disease in marine fishes. In: Snieszko SF (ed) A symposium on diseases of fishes and shellfishes. American Fisheries Society. Spec Publ No 5, Washington DC

Bagge J, Bagge O (1956) *Vibrio anguillarum* as the cause of ulcer disease in torsk (*Gadus callarias* Linné) (Danish). Nord Veterinaermed 8:481–492

Baross JA, Liston J, Morita RY (1978) Ecological relationship between *V. parahaemolyticus* and agar digesting vibrios as evidenced by bacteriophage susceptibility patterns. Appl Environ Microbiol 36:500–505

Cawley GD (1976) Some aspects of gill disease in farmed rainbow trout *(Salmo gairdneri)*. MSc Thesis. Stirling University

Cisar JO, Fryer JL (1969) An epizootic of vibriosis in chinook salmon. Bull Wildl Dis Assoc 5:73–76

Crosa JH, Schiewe MH, Falkow S (1977) Evidence for plasmid contribution to the virulence of the fish pathogen *Vibrio anguillarum*. Infect Immun 18:509–513

Evelyn TPT, McDermott LA (1961) Bacteriological studies of freshwater fish. I. Isolation of aerobic bacteria from several species of Ontario fish. Can J Microbiol 7:375–382

Harrell LW, Novotny AJ, Schiewe MH, Hodgins HO (1976) Isolation and description of two vibrios pathogenic to Pacific salmon in Puget Sound, Washington. US Fish Wildl Serv Fish Bull 74:447–449

Hodgkiss W, Shewan JM (1950) Pseudomonas infection in a plaice. J Pathol Bacteriol 62:655–657

Horne MT, Richards RH, Roberts RJ, Smith PC (1977) Peracute vibriosis in juvenile turbot *Scophthalmus maximus*. J Fish Biol 11:335–361

Horsley RW (1973) The bacterial flora of the Atlantic salmon *(Salmo salar* L) in relation to its environment. J Appl Bacteriol 36:377–386

Kusuda R (1966) Studies on the ulcer disease of marine fishes. US-Jpn Conf Mar Bacteriol, Tokyo, Aug 1966

Levin MA, Wolke RE, Cabelli VJ (1972) *Vibrio anguillarum* as a cause of disease in winter flounder *(Pseudopleuronectes americanus)*. Can J Microbiol 13:405–412

McArdle JF (1973) Studies on toxicity of *Vibrio anguillarum*. MSc Thesis. Stirling University

Richards RH, Pickering AD (1979) Changes in serum parameters of *Saprolegnia*-infected brown trout. *Salmo trutta* L. J Fish Dis 2:197–206

Rucker RR (1959) Vibrio infection among marine and freshwater fish. Prog Fish Cult 21:22–25

Tanaka J (1974) Vibrio infection of marine fish. Proceedings of 3rd US-Jpn Meeting Aquacult Spec Publ Jpn Fisheries Agencies and Sea Regional Fish Res Lab, Migato, Jpn, pp 113–114

Umbreit TH, Tripp MR (1975) Characterisation of the factors responsible for death of fish infected with *Vibrio anguillarum*. Can J Microbiol 21:1272–1274

Wood JW (1974) Diseases of Pacific salmon, their prevention and treatment (2nd ed). Washington State Department of Fisheries, Olympia, Washington, 82 pp

Experimental and Naturally Occurring Furunculosis in Various Fish Species: a Comparative Study

D. BUCKE[1]

Introduction

It has been known for many years that furunculosis affects a wide variety of freshwater fish. Plehn (1911) not only reported the disease in various teleosts, but also described experimental procedures for infecting certain of these species with furunculosis. Williamson (1929) experimentally transmitted the disease to the minnow (*Phoxinus phoxinus* L.) and the goldfish (*Carassius auratus* L.). Mawdesley-Thomas (1969) reported naturally-occurring furunculosis in the goldfish and described its histopathology.

Bootsma et al. (1977) suggested that *Aeromonas salmonicida*, the causative agent of furunculosis, was also the causative agent of the disease carp erythrodermatitis. McCarthy (1975) considered that a non-pigmented form of *A salmonicida* (var. *achromogenes*) was responsible for a large mortality of non-salmonids and, more recently, Bucke (1979) described the isolation of *A. salmonicida* from an epizootic specific for perch (*Perca fluviatilis* L.).

In this report a series of experiments is described which involved infection of brown trout and certain non-salmonid species with strains of *A. salmonicida*, and the histopathology resulting from these experiments is compared with the histopathology of natural infections of *A. salmonicida* in both salmonids and non-salmonids.

Materials and Methods

Experimental Procedures

The Fish

Six species of fish, brown trout (*Salmo trutta* L.), roach (*Rutilus rutilus* L.), rudd (*Scardinius erythrophthalmus* L.), carp (*Cyprinus carpio* L.), goldfish (*Carassius auratus* L.), and perch (*Perca fluviatilis* L.) were acclimatized to top water at a temperature of 16°–18°C. The fish were all within 7–14 cm body length.

---

[1] Ministry of Agriculture, Fisheries and Food, Directorate of Fisheries Research, Fish Diseases Laboratory, Weymouth DT4 8UB, Dorset, England

The Test Organisms

Three strains of *A. salmonicida*:
1) 18/77, pigmented strain isolated from a brown trout epizootic
2) 170/76, weakly-pigmented strain isolated from goldfish donated by Tavolek Inc. USA
3) 3/76, non-pigmented strain isolated from cyprinid fish epizootic.

Methods of Infecting Fish

1) Injection: six fish of each species were anaesthetised in MS222 and an intra-muscular (IM) injection of the test organism was given (0.1–0.25 ml, depending on body weight). This method was employed for all three strains of bacteria.
2) Cohabitation: three fish of each of the six species were placed by species in separate tanks containing 2 brown trout which had been injected IM with 0.1 ml of 1.4×10^6 of the isolate 170/76.

Controls

Three fish of each species were anaesthetised and given an IM injection of 0.1–0.2 ml N saline, and a further 3 fish of each species were left untreated.

The experiments were conducted in tanks which had a continous flow of tap water, plus individual aeration and a constant temperature of 16°–18°C. Each test was scheduled to last for 24 days, after which samples were taken from lesions and kidneys for bacteriological examination (after McCarthy, 1976) and from viscera, gills, muscle and injection sites for histological examination (after Bucke, 1972). The tests were checked daily, and samples from moribund fish were taken for both bacteriological and histological examination as they appeared. Dead fish were macroscopically examined only

Natural Infection

Isolates of *A. salmonicida* had been isolated during routine diagnostic examinations of epizootics of wild fish, including salmonids. The histopathological changes from some of these species were compared with those from the experimental infections described above.

Moribund fish were given an intramuscular injection of about 20 mg/kg of a corticosteroid, prednisolone acetate (PA, "Deltastab", Boots Ltd., England), and retained at 16°–18°C for 5 days (after McCarthy, 1978).

Results

Inoculation Experiment

Table 1 shows the variation and latent period of disease after inoculation, and as can be seen, there was variation between the species and the virulence of the isolates, although the trout was the most susceptible species to all three isolates.

Table 1. Latency of disease and severity of inoculation site lesions

	18/77	Isolates 170/76	3/76	Controls
Brown trout	< 5***	<5***	< 8***	24 days
Roach	< 9***	<5***	<24**	24*
Rudd	<11*	<5***	<24*	24
Carp	24**	<5**	24*	24
Goldfish	<24*	<5***	<24*	24*
Perch	<24*	<6*	<24***	24

<5, period of latency in days (fish were either dead or killed within this period); *** marked raised lesion; ** moderate lesion, ulcerated and exudate; * minimal lesion

The inoculation site lesions were present as raised oedematous areas, sometimes showing exudate, and in one species, the roach, fish became covered in mucous. Incised lesions revealed purulent fluid, and where the fish remained alive for a longer period there were extensive haemorrhagic infiltrations into the adjacent musculocutaneous tissues. Internal changes were not always obvious, but in some cyprinid species there was occasionally about 5 ml of pale haemorrhagic fluid present.

In all species inoculated with the bacterial isolates, histopathological changes were present in the form of necrosis associated with the injection sites. The control fish injected with saline also showed injection-site tissue reactions, but these were localised and did not reveal necrosis or hyperaemia. No significant pathological changes were seen in the control fish apart from small injection-site wounds and parasitic infestations which were common to all non-salmonid species.

Brown Trout

The inoculated brown trout showed marked injection-site reactions, mostly severe muscle necrosis, hyperaemia and serous exudate in the sinuses. Macrophages and occasional foci of lymphocytes were abundant in these necrotic areas, as were micro-colonies of bacteria, sometimes within muscle bundles. There was an increase in cellular activity in visceral organs, including the heart, with micro-colonies of bacteria present in heart, spleen and kidneys. There were also bacterial colonies present in the gill epithelia.

Non-Salmonids

Severe injection-site reactions showing micro-colonies of bacteria were prominent in the non-salmonid species. The changes in some species were more marked and spread over wider areas. In particular, perch, goldfish and carp species showed marked necrosis, serous exudate and hyperaemia.

The cyprinid species (roach, rudd, goldfish, and carp) all showed increases in cellular activity, and micro-colonies of bacteria were often spread along the sinusoids of the dermis.

Changes in visceral organs varied within the species. Cellular infiltration of the intestinal mucosa was common to all but no bacteria were observed in this site. Myocarditis was common to all species, especially fish inoculated with the isolate 170/76. Bacteria were present in the kidney interstitial tissues, gills and hearts of goldfish, carp and perch inoculated with this same isolate. Circulating mononuclear cells were commonly seen. Hyperplasia of the gill epithelium of the cyprinid species was more pronounced than in the control fish; however, infestation with *Ichthyophthirius* spp. was widespread in these species.

Cohabiting Experiment

The inoculated trout developed disease symptoms as reported above. In the test species only minimal changes were observed. Superficial haemorrhages and scale loss, with lepidorthoses, were typical in the cyprinid species, but not in perch. Although some fish died during the trial, the majority were killed at the termination time.

The histopathological changes included the presence of: (a) multi-nucleated cells in the peripheral myocardium in trout; (b) marked numbers of circulating mononuclear cells in roach; (c) marked cellular infiltration of visceral tissues, with evidence of necrosis in some organs in carp and goldfish. *A. salmonicida* was isolated from trout, goldfish, carp, and roach in this experiment.

Natural Infections

A. salmonicida was isolated from brown trout, rainbow trout (*Salmo gairdneri* Richardson), Atlantic salmon (*Salmo salar* L.), brook trout (*Salvelinus fontinalis* Mitchill), grayling *(Thymallus thymallus)*, perch (*Perca fluviatilis* L.), dace (*Leuciscus leuciscus* L.), bream (*Abramis brama* L.), roach, and goldfish. Typical furunculosis disease was often evident in the salmonids (Ferguson and McCarthy, 1978), but clinical and histopathological changes in other species were varied. Most of these isolates resulted from heavy mortalities, where the most usual signs were deep ulcerative skin lesions, sometimes only affecting one species. In these non-salmonid species, micro-colonies of bacteria so often seen in furunculosis in salmonids were not evident, although sites of bacteria were occasionally present in the necrotic muscle tissue. The visceral organs were infiltrated with mononuclear cells, and necrotic areas were observed occasionally.

In some instances, *A. salmonicida* was isolated only after moribund fish were injected with a corticosteroid. There were many instances when non-salmonid fish presented "typical symptoms of carp erythrodermatitis" (CE) (Bootsma et al., 1977) or other ulcerative conditions, but *A. salmonicida* was not isolated; however, the histopathological changes were similar to those previously described.

Discussion

It can be seen from the results described here that the sub-acute form of furunculosis can be reproduced by intra-dermal injections of *A. salmonicida;* Klontz et al. (1966)

demonstrated this in salmonids; in this study other species were also shown to be susceptible. It was possible to transmit the bacterium *A. salmonicida* by cohabiting healthy and diseased fish, although major pathological changes were not observed. However, when wild fish mortalities were investigated, the bacterium was isolated only occasionally, and then only after subjecting the fish to stress, even though clinical and histopathological signs of furunculosis or carp erythrodermatitis were present.

This raises two points:

1. It is important to know whether *A. salmonicida* is the causative agent of the many ulcerative diseases of non-salmonids, especially in perch, a species in which there is an endemic disease in the UK (Bucke et al., 1979). Despite intensive investigations, *A. salmonicida* has only occasionally been isolated and no other significant organisms have been identified, yet the disease persists.

2. It is also important to know the significance of *A. salmonicida* in non-salmonids, because of the potential reservoir it would represent of a dangerous pathogen of farmed salmonids.

The reference to proprietary products in this report should not be construed as an official endorsement of these products, nor is any criticism implied of similar products which have not been mentioned.

References

Bootsma R, Fijan N, Blommaert J (1977) Isolation and preliminary identification of the causative agent of carp erythrodermatitis. Vet Arch 47:291–302

Bucke D (1972) Some histological techniques applicable to fish tissues. Symp Zool Soc London 30:153–189

Bucke D (1979) Investigation into the cause of an epizootic in perch (*Perca fluviatilis* L.). In: O'Hara K (ed) Proc 1st Br Freshwat Fish Conf. Univ Liverpool. Janssen Services, London, p 45

Bucke D, Cawley GD, Craig JF, Pickering AD, Willoughby LG (1979) Further studies of an epizootic of perch *Perca fluviatilis* L., of uncertain aetiology. J Fish Dis 2:297–311

Ferguson HW, McCarthy DH (1978) Histopathology of furunculosis in brown trout *Salmo trutta* L. J Fish Dis 1:165–174

Klontz GW, Yasutake WT, John Ross A (1966) Bacterial disease of the Salmonidae in the Western United States: Pathogenesis of furunculosis in rainbow trout. Am J Vet Res 27:1455–1460

Mawdesley-Thomas LE (1969) Furunculosis in the goldfish, *Carassius auratus* (L.). J Fish Biol 1:19–23

McCarthy DH (1975) Fish furunculosis caused by *Aeromonas salmonicida* var. *achromogenes*. J Wildl Dis 11:489–493

McCarthy DH (1976) Laboratory techniques for the diagnosis of fish furunculosis and whirling disease. Fish Res Tech Rep MAFF Direct Fish Res Lowestoft, No 23, p 5

McCarthy DH (1978) Some ecological aspects of the bacterial fish pathogen, *Aeromonas salmonicida*. In: Aquatic microbiology, Soc Appl Bacteriol Symp No 6, p 299

Plehn M (1911) Die Furunkulose der Salmoniden. Zentralbl Bakteriol Parasitenkd Abt I 60:609–624

Williamson IJF (1929) A study of bacterial infection in fish and certain lower vertebrates. In: Salm Fish Edinb No 11:3

Seasonal Occurrence of Aeromonas salmonicida Carriers

N. J. JENSEN and J. L. LARSEN[1]

Introduction

Furunculosis is one of the most economically important bacterial diseases among salmonids in fish farms, and is very often controlled by use of antibiotics and chemotherapeutics. Sulfamerazine has generally been the drug of choice (Snieszko, 1978).

From an environmental quality point of view it is not advisable to use antibiotics and chemoterapeutics routinely in the natural environment as antibiotic-resistant bacteria might emerge. This problem also implies aspects of interest for human public health.

Attempts to control furunculosis by means of vaccines were made by Antipa and Amend (1977) and Udey and Fryer (1978), and their results indicate that it is a possible method. However, further investigations must be carried out before an effective prophylaxis can be established. The elucidation of the interaction between host, pathogen, and environment in cases of nonclinical furunculosis may contribute to a better understanding of the pathogenesis of the disease and the development of preventive measures. Bullock and Stuckey (1975) developed a method for the detection of symptom-free carrier of *Aeromonas salmonicida*. This method was later used in different modifications by Jensen (1977) and McCarthy (1977). To study the host-parasite relationship a seasonal registration of the carrier rate of *Aeromonas salmonicida* was performed.

Materials and Methods

Fish from two fish farms which usually had problems with furunculosis were selected for a seasonal study of the carrier prevalence. One of the fish farms cultured mainly brown trout *(Salmo trutta)* and the other rainbow trout *(Salmo gairdneri)*.

The carrier detection experiments were carried out according to Jensen (1977). Water temperature and oxygen content in the water were measured by an EIL oxygen meter 1520 at the time the fish were taken from the fish farm.

---

1 Royal Veterinary and Agricultural University, Copenhagen, Denmark

Table 1. Seasonal carrier rate of *Aeromonas salmonicida* in brown trout and rainbow trout from two different trout farms

Fish	Brown trout		Rainbow trout	
Water temperature at the fish farm	17 °C	2 °C	15 °C	1 °C
Oxygen content in the water at the fish farm	50%–60%	90%–100%	45%–55%	90%–100%
Experimental water temperature	18 °C	18 °C	23 °C	23 °C
Number of fish	20	46	45	56
Number of fish infected	20	12	38	3
Infected with *A. salmonicida*	18	0	25	0
Infected with *A. salmonicida* (%)	90	0	66	0
Infected with *A. hydrophila*	2	12	13	3
Infected with *A. hydrophila* (%)	10	26	34	5
A. salmonicida carrier rate (%)	90	0	55	0

Results

The seasonal prevalence of the *Aeromonas salmonicida* carriers is listed in Table 1.

The brown trout generally died within 3–6 days in the summer period, but in the winter period half of the brown trout were alive on the 6th day. Most of the rainbow trout survived the 6th day in the winter period. The trouts were subsequently killed and submitted to a bacteriological examination.

Discussion and Conclusion

A simultaneous challenge of brown trout *(Salmo trutta)* and rainbow trout *(Salmo gairdneri)* to the stress of corticosteroids and elevated temperature showed a very characteristic pattern in the seasonal occurrence of *Aeromonas salmonicida* carriers in farms with furunculosis problems.

These carriers were restricted to the summer period, whilst the *Aeromonas hydrophila* infection was found both in the summer and the winter period. There is a clear and important difference in the biology of the two *Aeromonas* species. *Aeromonas salmonicida* seems to be a pathogen generally attached to the trouts, while *Aeromo-*

nas hydrophila is a bacterium with habitat in aquatic environments, where the prevalence is influenced by temperature and the presence of organic matter (pollution).

In general the *Aeromonas salmonicida* carriers occur concomitant with the appearance of clinical furunculosis in the locality in question. The appearance of the disease is greatly influenced by high water temperature and it seems therefore to disappear during the winter period, when no carriers could be found.

In spring when trout are transferred to other fish farms or a marine environment, the problem of the existence of carriers arises. These carriers could bring *Aeromonas salmonicida* into the environment. When the temperature rises in early summer outbreaks of furunculosis occur in fish with another immunological status. It is therefore important to develop methods to find such nondetectable carriers, and disclose the real habitat of the bacterium during the winter. Such information could be used as guidance on prophylactic measures to be used, such as introduction of *Aeromonas-salmonicida*-free trout or application of a vaccination method.

References

Antipa R, Amend DF (1977) Immunization of Pacific salmon and comparison of intraperitoneal injection and hyperosmotic infiltration of Vibrio anguillarum and Aeromonas salmonicida bacterins. J Fish Res Board Can 34:203–208

Bullock GL, Stuckey HH (1975) Aeromonas salmonicida: Detection of a symptomatically infected trout. Prog Fish Cult 37(4):237–239

Jensen NJ (1977) The diagnosis of furunculosis in salmonids. Bull Off Int Epiz 87(5–6):469–473

McCarthy DH (1977) Some ecological aspects of the bacterial fish pathogen. In: Skinner FA, Sherwan JM (eds) Aeromonas salmonicida in Aquatic microbiology. Academic Press, London New York, pp 299–324

Snieszko SF (1978) Control of fish diseases. Mar Fish Rev 40(3):65–68

Udey LR, Fryer JL (1978) Immunization of fish with bacterins of Aeromonas salmonicida. Mar Fish Rev 40(3):12–17

Examination in Resistance Tests of Some Strains of the Aeromonas hydrophila punctata Group Isolated from Carp

W. NEUMANN and W. PLÖGER[1]

The application of antibiotics is a long-known and practicable method to control bacterial diseases of fish. It is in the interests of an effective therapy to observe constantly the sensibility of bacterial organisms.

In our investigations started in 1970 strains of the *Aeromonas hydrophila punctata* group isolated from carp of Northwest German fish farms were examined in resistance tests. In regard to its secondary influence on the erythrodermatitis of carp caused by *A. salmonicida* (Fijan, 1978) the bacterial group mentioned is of great interest.

Materials and Methods

The present investigations refer to 60 *Aeromonas* strains isolated from livers and kidneys of carp in the years 1970 to 1978 with the following common characteristics:
a) Microscopically: short gram-negative rods
b) Culture: haemolysis on bovine blood agar; yellow colonies on GSP-Agar (Merck, Darmstadt); red colonies on brillant-green phenol-red agar
c) Biochemistry: reduction of glucosis under aerobic and unaerobic conditions; positive cytochrome oxidase reaction; turbidity of peptone water.

The sensitivity test was executed on DST-Agar (Oxid) under addition of 7% bovine blood. Mastring-S test stars (Mast-Diagnostika, Hamburg) served as antibiotic carriers. The technique was described by Neumann and Plöger (1975). Qualities and quantities of the antibiotic substances are demonstrated in Tables 1 and 2.

Results and Discussion

First we compared those antibiotics which are generally components of commercial feeding drugs such as Chloramphenicole, Chlortetracycline, Furazolidone, and Trimethoprim/Sulfadimethoxin (TMP/S). According to these investigations we could not determine an increase of resistance of *Aeromonas hydrophila punctata* strains to Chloramphenicole, Chlortetracycline and Furazolidone within the last 9 years. Sensitivity could be confirmed against Chloramphenicole in 56, against Chlortetracycline

---

[1] Tiergesundheitsamt der Landwirtschaftskammer Weser-Ems, Postfach 2549, D-2900 Oldenburg, FRG

Table 1. Numerical presentation of efficacy of antibiotics and chemotherapeutics

Year	No of strains	Chloramphenicole			Chlortetracycline			Flurazolidone			TMP/S		
		E	e	R	E	e	R	E	e	R	E	e	R
1970	1	1	–	–	–	1	–	–	–	1	Not tested		
1971	3	2	1	–	2	1	–	–	3	–			
1972	8	4	1	3	2	4	2	6	2	–			
1973	2	1	1	–	–	2	–	–	2	–			
1974	6	6	–	–	2	2	2	3	3	–	2	1	3
1975	3	2	1	–	2	1	–	1	2	–	3	–	–
1976	8	6	2	–	3	5	–	5	3	–	2	3	3
1977	14	9	5	–	7	7	–	11	3	–	3	6	5
1978	15	9	5	1	8	6	1	9	5	1	2	7	6
Total	60	40	16	4	26	29	5	35	23	2	12	17	17

Test star charges:
Chloramphenicole   10 μg (from 1975 50 μg)   Flurazolidone   100 μg
Chlortetracycline   10 μg (from 1975 50 μg)   TMR/S   25 μg
E, highly sensitive
e, slightly sensitive
R, not sensitive

in 55, and against Furazolidone in 58 strains from the total of 60 isolates. Only TMP/S proved less activity, showing resistance in 17 isolates (=28%).

Moreover and in accordance with the results mentioned above we found that 10 strains selected at random presented high sensibility to the overwhelming majority of 16 different antibiotics. There was no *Aeromonas* strain showing resistance to all of the 16 drugs. No efficacy could be ascertained with Penicillin and Cloxacillin. Also Sulfadimidin and Spiramycin did not satisfy the requirements.

The results of our investigations correspond to those of Krabisch (1976), who examined *Aeromonas* strains from carp in Bavaria. Little differences may be due to some unequal charges of the antibiotic test stars. By the way, the above-mentioned substances seem to possess good efficacy also against *A. salmonicida* as described by Krabisch and Wiedemann (1979).

Considering that every in vitro test will never imitate the natural circumstances completely the Antibiogram nevertheless represents a useful support in effective treatment of bacterial diseases of fish. Practical success and experience confirm this statement.

Although the present situation of resistance of *Aeromonas* species to antibiotics seems to be statisfactory in regard to therapy, the further developments should be observed attentively and critically.

Table 2. Antibiogram of some representative *Aeromonas* strains of the *Hydrophila punctata* group

Strain No	Chloramphenicol 50 µg	Chlortetracycline 50 µg	Erythromycin 50 µg	Oxytetracycline 50 µg	Gentamycin 10 µg	Penicillin-G 10 E	Neomycin 30 µg	TMP/S 25 µg	Streptomycin 25 µg	Colistin 10 µg	Cloxacillin 10 µg	Rifampicin 30 µg	Tylosin 30 µg	Spiramycin 100 µg	Sulfadimidin 500 µg	Furazolidone 100 µg	R	E	e
90/75	E	E	e	E	E	R	E	E	E	R	R	E	e	e	R	e	4	8	4
579/75	E	E	E	E	E	R	E	E	e	e	R	E	e	e	R	E	3	9	4
632/75	e	e	E	e	E	R	E	E	E	R	R	E	e	e	R	e	4	6	6
1849/75	E	E	e	E	E	R	e	e	e	e	R	e	e	e	R	e	3	4	9
1005/76	e	E	E	E	E	R	E	e	e	E	R	E	R	e	R	E	4	9	3
1020/76	E	e	E	e	E	R	E	R	R	R	R	e	R	e	R	e	7	3	6
1029/76	E	e	e	e	E	R	e	E	E	e	R	e	R	e	e	E	5	4	7
1060/76	E	E	E	E	E	R	E	E	e	e	R	E	e	e	R	E	2	9	5
1412/76	E	e	E	R	E	R	E	e	e	e	R	e	e	e	R	e	4	4	8
1463/76	E	e	E	e	E	R	E	e	e	e	R	E	e	e	R	e	3	5	8
R	0	0	0	1	0	10	0	2	1	3	10	0	3	0	9	0			
E	8	5	7	5	10	0	8	4	3	1	0	6	0	0	0	4			
e	2	5	3	4	0	0	2	4	6	6	0	4	7	10	1	6			

R, not sensitive
E, highly sensitive
e, slightly sensitive

References

Fijan N (1978) Dtsch Veterinaermed Ges, Fachgruppe Tierseuchenrecht, Tagungsbericht München 24.–26.10.1978, S 125–131
Krabisch P (1976) Fisch und Umwelt 2:119–120
Krabisch P, Wiedemann H (1979) Tierärztl Prax 7:81–90
Neumann W, Plöger W (1975) Dtsch Tierärztl Wochenschr 82:80–83

Development of Bacteria in Fish and in Water During a Standardized Experimental Infection of Rainbow Trout (Salmo gairdneri) with Aeromonas salmonicida

C. MICHEL[1]

In the opinion of many workers, one of the major difficulties in studying furunculosis of salmonids has been the experimental reproduction of the disease (Amend, 1969; Anderson, 1972). Notwithstanding some success obtained with parenteral injection of low doses of the responsible agent, *Aeromonas salmonicida* (Groberg et al., 1978), this problem has never been well solved. We recently described a standardized method using intramuscular injection of highly virulent bacteria to rainbow trout *(Salmo gairdneri),* which has given consistent results and has proven fairly reliable in testing the efficacy of therapeutic treatments (submitted for publication). The difficulties that we had to overcome during this work and the proposed solutions are summarized in Table 1. In the present work we have tried to give some additional features of this model, concerning the fate of the causative organism, both when it multiplies in the fish and when it is shed in water by infected fish. Especially in the latter case, serious hazards of superinfections could have interfered with standardization procedure.

Table 1. Main difficulties encountered in developing the experimental infection with *A. salmonicida*

Problems	Criteria for appreciation	Solutions
1. Selection and maintenance of *virulence*	Low LD_{50}	Serial passages of strains on rainbow trout fingerlings
2. Choice of a *penetration route*	Efficacy Regularity of mortality curves	IM injection
3. Control of administered *doses*	Linear relation between OD and number of viable cells in bacterial suspensions	Early harvesting of stirred cultures (OD ≤ 0.800)
4. *Standardization*	Reproducibility of experiments	Control of experimental conditions Batches divided into randomly distributed subgroups

[1] Laboratoire d'Ichtyopathologie, Station de Virologie et d'Immunologie, 78850 Thiverval-Grignon, France

Material and Methods

All fish were held in 12-l aquaria supplied with flowing dechlorinated tap water. The temperature was 15 °C, and the flow rate about 15 l/h.

Rainbow trout fingerlings originating from a furunculosis-free hatchery in Normandy (mean weight 18 g) were used in all experiments. Lack of antifurunculosis agglutinins was checked on a sample of 15 fish before experimentation.

A virulent strain of *A. salmonicida*, TG 36/75, was used. It is a self-aggregating form, and its lethal dose 50 per 100 (LD_{50}) was about 5,000 bacteria per fish, IM. This strain is routinely subcultivated in fish, and the procedure for preparing infectious suspensions has been described: stirred cultures of bacteria in tryptic soy broth at 22 °C were harvested at a very early state of growth, and the infective doses adjusted in spectrophotometry at 525 nm. Infections were performed in conformity to the described standard, after anesthesia with MS 222 (Sandoz) at 50 ppm.

In a first experiment, 30 fingerlings were acclimatized in an aquarium and infected with 10,000 bacteria per fish. In order to follow the course of the disease, 3 trout were killed every day and blood, liver, and kidney were plated on tryptic soy agar for isolation. Similar processes were also applied to fish dying from the disease.

The second experiment involved 3 aquaria. The first one was supplied with 10 noninfected controls. The two others each received 15 clipped opercula trout respectively inoculated with 5,000 and 50,000 bacteria, and 15 nonbranded and healthy trout which served to detect possible superinfections. Mortality was recorded, and in addition bacteriological controls of water were made every day, in order to estimate the total microbial population of the aquaria and the importance of *A. salmonicida* release. For this purpose we used the Miles and Misra (1938) droplets enumeration technique, on TSA medium: each water sample was diluted tenfold. Discrimination of *A. salmonicida* colonies was easy when reading the plates under 45 ° transmitted light.

Results and Discussion

No fish survived the first experiment (Table 2). The first mortalities occurred on the 4th day, and all fish had died on the 7th, with maximum loss on the 5th day. So, with an infective dose of 2 LD_{50}, the disease has a rapid course, which results in septicemia: hemorrhagic lesions were observed, and the *Aeromonas* was always isolated from internal organs. During the first days, a bump appeared at the site of injection and matured as a typical "furuncle" which usually opened on the 4th day. Table 2 shows that for 2 days we failed to detect the germ in blood flow and hematopoietic organs of killed animals. It is only on the 3rd day, 24 h before recording the first mortalities, that the bacteria could be isolated from internal organs, even in fish which displayed no general sign of disease. It seems consequently that a local multiplication of the germ of about 3 days precedes a systemic infection which results very quickly in death. Such an evolution looks like the subacute form of furunculosis described by McCraw (1952). It is of interest to note that furuncles mature only at the injection site. They seem to correspond to a very local multiplication in muscle and perhaps, in natural conditions, their occurrence could reveal penetration points of the pathogen through the skin.

Table 2. Course of furunculosis infection after IM injection of 10,000 bacteria in 18-g rainbow trout fingerlings

Days after infection	Mortality due to the disease	Killed fish	Frequency of isolation of germs on killed fish		
			Blood	Liver	Kidney
1	0	3	0	0	0
2	0	3	0	0	0
3	0	3[a]	2	2	2
4	2	2[b]	2	2	2
5	12	0			
6	4	0			
7	1	0			

[a] Only 1 fish exhibited signs of disease
[b] 1 fish exhibited signs of disease

The development of the bacterial population in water is shown in Table 3. The total number of bacteria ranged generally from 10,000 to 40,000 cells/ml. An interruption of water supply occurred on the 7th day for 4 h, increasing these numbers considerably. It can be seen that *A. salmonicida* appeared in water only between 48 and 72 h after injections, following the maturation of furuncles, and persisted with irregular variations until about 48 h after removal of the last dead trout. Consequently it could be expected that superinfections were possible and could hinder the model standardization, unless a latent period exists in such cases, which could be longer than the observation time of 10 days chosen for the model. This seems to be verified if one refers to Table 4. After 10 days, no fish among noninjected controls had died in spite of the presence of virulent bacteria in the water.

Moreover, superinfections do not occur in all cases, and do not appear related to the quantity of bacteria released in the environment. Aquarium 3 had shown highter

Table 3. Development of the total microbial population and of *A. salmonicida* in water containing 15 experimentally infected rainbow trout fingerlings (trout weight 15–18 g; $T°=15\ °C$)

Days after infection	Aquarium 1 (infective dose 5,000 b IM)			Aquarium 2 (infective dose 50,000 b IM)		
	Mortality	Bacteria (10^3/ml)	*A. salmonicida* (10^3/ml)	Mortality	Bacteria (10^3/ml)	*A. salmonicida* (10^3/ml)
1	0	5	0	0	1.5	0
2	0	10.5	0	0	2	0.15
3	0	45	5.5	0	10	5.5
4	1	19.5	1.5	3	17.5	7
5	2	30	4	7	14.5	1.5
6	3	9.5	2.5	0	8	2
7	1	120	2	1	120	60
8	0	40	0.03	1	14.5	6
9	0	7	0.15	1	18	6.5
10	0	22.5	0	0	11	2.5

Table 4. Transmission experiment from furunculosis-infected to furunculosis-free fingerlings (rainbow trout 15–18 g; *A. Salmonicida* strain TG 36/75 "R"; $LD_{50} \simeq 5{,}000$ b/fish IM); $T° = 15\,°C$)

Aquarium	Fingerlings	Numbers	Mortalities at D_{10}	Mortalities at D_{30}
1	Noninfected controls	10	0	0
2	Injected (5,000 b IM)	15	7	9
	Noninjected	15	0	2
3	Injected (50,000 b IM)	15	13	14
	Noninjected	15	0	0

titers of *A. salmonicida* than the others (the inoculated fish had received the largest dose). Especially during the water supply break these titers had grown up to 60,000 bacteria/ml! If one considers that a fish dead on day 11 probably succumbed to the primary injection, it is only in the other batch that a chronic furunculosis broke out between the 10th and 30th days, involving previously injected subjects as well as noninjected ones.

In conclusion, our experimental model of furunculosis can be used without risk of superinfections, and the standardized schedule appears reliable. It simulates a subacute form of the disease, and certainly this feature limits its value: other forms of the disease are known in nature, and the mechanisms of pathogenicity and host reaction involved probably vary. However, the method should prove useful, allowing us to study at least partially this kind of problem.

References

Amend DF (1969) Oxytetracycline efficacy as a treatment for furunculosis in coho salmon. US Bur Sport Fish Wildl, Tech Pap 31, p 6

Anderson DP (1972) Virulence and persistance of rough and smooth forms of *Aeromonas salmonicida* inoculated into coho salmon *(Oncorhynchus kisutch)*. J Fish Res Board Can 29:204–206

Groberg WJ, McCoy RH, Pilcher KS, Fryer JL (1978) Relation of water temperature to infections of coho salmon *(Oncorhynchus kisutch)*, chinook salmon *(O. tshawytscha)*, and steelhead trout *(Salmo Gairdneri)* with *Aeromonas salmonicida* and *A. hydrophila*. J Fish Res Board Can 35:1–7

McCraw BM (1952) Furunculosis of fish. US Fish Wildl Serv Spec Sci Rep Fish 84:87 p

Miles AA, Misra SS (1938) The estimation of the bacteriocidal power of the blood. J Hyg 38:732–749

Studies on an Ichthyotoxic Material Produced Extracellularly by the Furunculosis Bacterium Aeromonas salmonicida

A.L.S. MUNRO, T.S. HASTINGS, A.E. ELLIS, and J. LIVERSIDGE[1]

Introduction

The pathology of furunculosis shows considerable evidence of the production of virulence factors by the causative agent *Aeromonas salmonicida* (Mackie and Menzies, 1938; Klontz et al., 1966; Ferguson and McCarthy, 1978). Indeed there are accounts of haemolytic (Karlsson, 1962) and leucocytolytic activity (Klontz et al., 1966; Fuller et al., 1977) as well as protease (Shieh and McLean, 1975) production. Production of enzymes showing similar and other activites has been demonstrated in other Aeromonad species (Wadstrom et al., 1976). However, there has been little success in demonstrating that these extracellular products either singly or in combination can reproduce in part or, in whole, the pathology of furunculosis. Attempts with cell wall preparations of *A. salmonicida* have failed to show pathology in fish (Ross, 1966; Paterson, 1972; Anderson, 1973).

In this paper we describe a method of producing a lethal (for fish) extracellular product (ECP) from an isolate of *A. salmonicida* and some activities of this product in vivo and in vitro.

Methods

The isolate of *A. salmonicida* used throughout this study was obtained from a naturally occurring outbreak of furunculosis in Atlantic salmon smolts in Scotland.

ECP Production

The ECP of *A. salmonicida* was prepared using the cellophane overlay technique (Liu, 1957). It was prepared by harvesting the products of 48 h growth at 22 °C from several trays 30x24 cm containing Difco furunculosis agar overlaid with a sterile sheet of cellophane. The cells were washed off the cellophane with an equal volume of phosphate-buffered saline, then centrifuged, and the supernate (called ECP) was filtered (0.22 mμ) and kept in aliquots at -15 °C. Protein was determined by the method of Lowry et al. (1951), proteolysis by the method of McDonald and Chen (1965) as described by Shieh and MacLean (1975) and serum inhibition of protease by a method de-

---

[1] Marine Laboratory, Aberdeen Scotland, U.K.

scribed by Holder and Haidaris (1979). Extracellular protease and lecithinase activity of the culture of *A. salmonicida* was demonstrated using methods described by Cruickshank et al (1975).

Serology

Sera were obtained by bleeding anaesthetised fish, allowing a clot to develop overnight at 4 °C and withdrawing clear fluid from beneath the clot. Sera used in the haemolysis test were heated to 45 °C for 20 min to destroy naturally occurring haemolysins. The same sera were used for the *A. salmonicida* slide agglutination test. Immuno-electrophoresis was conducted using cellulose acetate strips (Cellogel, Whatman Ltd.) in 0.04 M barbitone sodium buffer pH 9.25 at room temperature and a constant 200 V for 30 min. Strips were stained in 0.005% nigrosine in 5% acetic acid.

Cell Preparations

Red blood cells (RBCs). Rainbow trout RBCs were collected in an equal volume of Alsever's solution and stored at 4 °C for at most 24 h. For use in haemolysis testing they were washed in either PBS or maintenance minimal essential medium (MMEM) and suspended at 1% v/v.

Leucocytes from the spleen of a 500-g rainbow trout were collected in silicone-treated glassware by the method of Bøyum (1968) using Histopaque (see Sigma Chemical Co. Bulletin No. 1077). Yields in excess of 10^8 leucocytes were routinely achieved with apparent viabilities (Trypan blue dye exclusion test) of >90%.

Macrophages were prepared by suspending the leucocyte preparation in growth minimal essential medium (GMEM) and adding this to a 90-mm petri dish containing clean, un-siliconised 10-mm glass coverslips. After 2 h the coverslips were removed by forceps and vigorously washed in 2 changes GMEM to remove non-adherent cells. The procedure yielded 50 coverslips each with 5–10 motile macrophages.

Rainbow-trout gonad (RTG-2) tissue culture cells were maintained on GMEM.

Test Procedure Using ECP and Cell Preparations

Microtitre plates containing wells in rows of 12x7 deep were used. 50 μl of RBCs in MMEM were added per well just prior to use. Leucocytes were seeded at 10^5 cells per well in 0.075 GMEM and incubated overnight at 15 °C. RTG-2 cells were seeded at 5×10^3 cells per well in 75 μl GMEM and incubated overnight at 20 °C reaching 50% confluence. Wells containing leucocytes and RTG-2 cells were rinsed to ensure removal of serum components of GMEM and then 50 μl of MMEM added. Twenty-four-hour-old coverslip macrophage cultures were rinsed and 4 coverslips placed in each of 12x30-mm petri dishes containing 2 ml of MMEM. Serial twofold dilutions (1/10 to 1/20,480) of ECP were prepared and 50 μl/ml added to 7 replicate rows of wells of RBCs, leucocytes and RTG-2 cells. 2 ml of each ECP dilution was added to the petri dish preparation of macrophages.

Cells were observed after 2 h exposure to ECP and recorded as lysis of all cell types. For other effects see results.

Test Procedure to Determine the Effect of Fish Serum on ECP Haemolytic Activity

100 μl aliquots of doubling dilutions of ECP were placed in microtitre wells. 50 μl of 1/20 dilution of serum was added and allowed contact for 20 min before addition of 50 μl of 2% v/v washed RBCs. Haemolysis was read after 30 min.

Pathology

The same ECP preparation was injected into rainbow trout. Preliminary experiments showed that IP injection of 0.5 ml killed fish in 2 h, and 0.2 ml killed fish (100 g) within 6 h, while IM injection produced a furuncle and eventual death as well. In order to allow time for a pathology to develop a sublethal dose of ECP was injected. Accordingly rainbow trout weighing 150 g were injected with 0.15 ml ECP. Six fish were injected IP and 5 fish by IM route in the dorsal musculature lateral to the dorsal fin. One IP-injected fish was moribund 11 h later and one IM-injected fish was found dead 30 h after injection. The remaining fish were killed for gross and histopathological examination 36 h after injection.

Results

Properties of the Isolate Used in this Study

This isolate of *A. salmonicida* in common with most cultures we have examined showed haemolytic activity on blood medium, proteolytic activity on casein and gelatine media and lecithinase activity on egg-yolk medium. The ECP showed haemolysis of fish and rabbit red cells, protease activity on casein and lecithinase activity on egg yolk. It contained 6.7 mg protein/ml and 18.7 units of specific protease activity/mg protein.

Effect of ECP Injected into Rainbow Trout

Fish injected IM with ECP developed a large swelling (furuncle) at the site of injection (Fig. 1) within 12 h. When lanced the furuncle was filled with a pink viscous fluid and the surrounding muscle showed signs of liquifaction and haemorrhage. Distal from the site of injection, liquifaction of the myotomal septa had occurred. The spleen was grossly enlarged and congested with blood and the liver was discoloured by dark red flecks.

Most fish injected by the IP route became lethargic and darker in colour soon after injection. They exhibited increased opercular activity and, while many recovered after several hours, one fish lost equilibrium and floated upside down apparently in a comatose condition until it was killed 11 h later. In this group the vent was inflamed

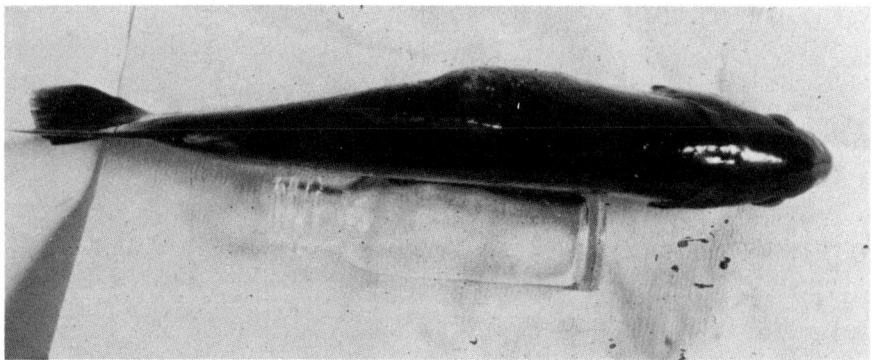

Fig. 1. Rainbow trout injected IM with the ECP of *A. salmonicida* showing an extensive swelling typical of the furuncle found in the chronic form of the disease, furunculosis

and the blood vessels in the rectum wall were frequently grossly dilated. The spleen was enlarged and occasionally the liver showed dark red flecks as observed in the IM-injected group. The peritoneal cavity often contained a small accumulation of ascitic fluid but otherwise no pathological features were apparent there.

A detailed account of the extensive histopathological changes occurring in brain, spleen, gut, gill, kidney, liver, and muscle is described by Ellis et al. (1981).

Effects of ECP on Host Cells and RTG-2 Cells

The consequences of exposing doubling dilutions of ECP to washed rainbow trout RBCs, to RBS-free preparations of washed leucocytes, macrophages and to RTG-2 tissue culture cells are shown in Table 1. Apparent differences in lytic titre and endpoint between different cell types are more apparent than real because beyond lysis endpoints other less dramatic effects were observed. For example, in the RBC preparation some release of haemoglobin was seen in the 3 wells to the right of the lytic endpoint and in all the leucocyte-containing wells to the right of lysis some proportion of cells had lysed. Cell rounding and retraction of pseudopodia were features of the effects on macrophages at ECP dilutions causing less than total lysis and in the RTG-2 cells, Trypan blue uptake in a proportion of cells 5 dilutions beyond total lysis could be seen.

Serum Inhibition of Haemolysis by ECP

When fish serum was reacted with ECP and the reaction mixture tested for haemolytic activity, considerable depression of ECP haemolytic activity was noted with all sera from three salmonid species (Table 2). There was little depression with rabbit serum. The *A. salmonicida* agglutination titre was very low in all the sera. Using the result in Table 2 it was calculated that 5 volumes of serum were required to inhibit haemolytic activity in 1 volume of the ECP preparation.

Table 1. Effects of exposing various cell populations for 2 h to doubling dilutions of the ECP of *A. salmonicida* (each row represents the mean of 7 rows)

Dilution of ECP	1/10→1/40	1/80	1/160	1/320	1/640	1/1,280	1/2,560	1/5,120	1/10,240	1/20,480
RBCs	L	L	L	L	L	L	N	N	N	N
Leucocytes	L	L	L	L	L	← Decreasing proportions of lysed cells <10% →		Some release of Haemoglobin →		N
Macrophages	L	L	L	L	L	← Cell rounding →	← Retraction of Pseudopodia →		N	N
RTG-2 cells	L	>25% Death ——— Trypan blue dye exclusion test ——— <10% Death					N	N	N	N

L, 100% cell lysis   N, no visible effect

Table 2. The effect of various fish and rabbit sera on the haemolysis endpoint of washed rainbow trout RBCs by *A. salmonicida* ECP

Well contents	Doubling dilution of ECP giving haemolysis	*A. salmonicida* cell agglutination titre
Rainbow trout RBC + ECP	1/4,128	—
Rainbow + rainbow trout serum (4 fish)	1/192	<1/2
Rainbow + salmon serum (4 fish)	1/176	<1/2
Rainbow + brown trout serum (4 fish)	1/608	<1/2
Rainbow + rabbit serum	1/2,048	<1/2

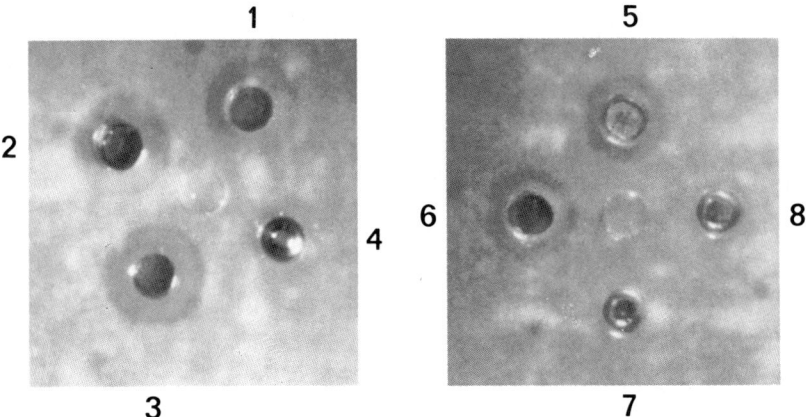

Fig. 2. Inhibition of protease activity by rainbow trout serum on a skim milk agar substrate. *1* ECP; *2* Trypsin; *3* 1 vol ECP to 1 vol serum; *4* 1 vol trypsin to 1 vol serum; *5* 1 vol ECP to 4 vol serum; *6* 1 vol ECP to 5 vol serum; *7* 1 vol ECP to 7 vol serum; *8* 1 vol ECP to 8 vol serum

Serum Inhibition of Protease Activity

Diffusion tests using skim milk protein in agarose were used to show that rainbow trout serum inactivated trypsin and ECP protease (Fig. 2). Undiluted serum inactivated a sevenfold dilution of ECP. The ECP produced a diffusing opaque ring in this test system but a zone of proteolysis can be seen at the leading edge of the opaque band.

Immuno-electrophoresis

Figure 3a shows that after electrophoresis both normal rainbow trout and rabbit serum produce a precipitin in the fast-moving α region when allowed to diffuse against a trough containing ECP. In Fig. 3b normal trout serum produces an almost identical precipitin against both ECP and trypsin in the α-migrating region. The precipitates were not soluble in citrate or ethylene diamine tetraacetic acid (EDTA) (Baldo and Fletcher, 1973).

Discussion

This report is the first record that the extracellular products of *A. salmonicida*, grown on cellophane overlay, reproduce most if not all of the pathology of furunculosis without the presence of the bacterium. Sufficient ECP given IP is rapidly fatal for fish. In lesser amounts it is also fatal but prolongs the time to death such that extensive histopathological changes occur in brain, spleen, intestine, heart, gill, liver, and kidney. The histopathology is reported elsewhere but corresponds in most respects with numerous descriptions of the disease. In addition the necrosis of muscle at the site of IM injection and the oedematous swelling there with collection of pink fluid plus

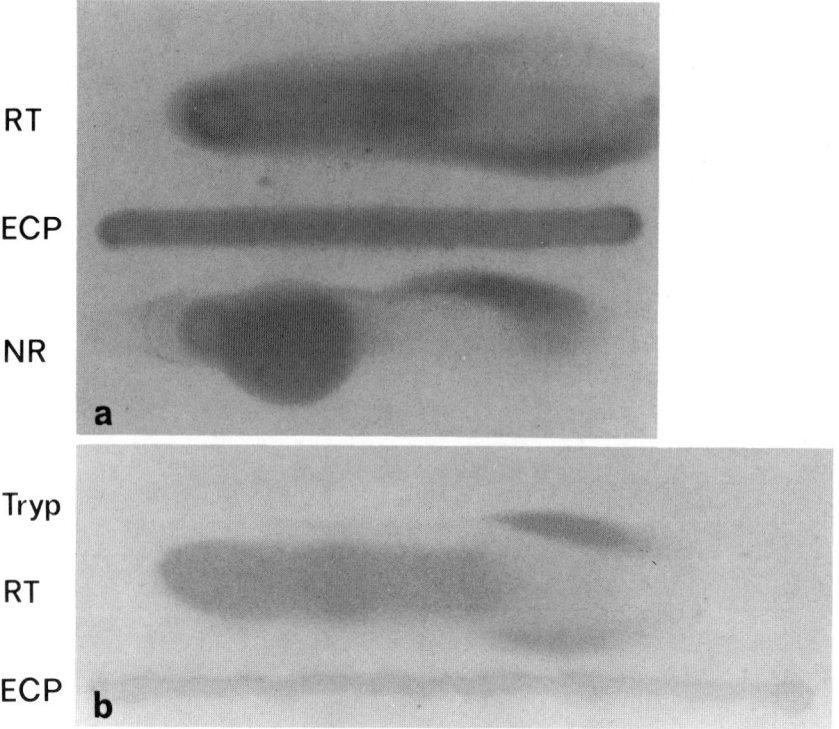

Fig. 3 a, b. Immunodiffusion tests. **a** Electrophoresis of rainbow trout serum *(RT)* and normal rabbit *(NR)* serum diffused against *A. salmonicida* ECP in centre trough. **b** Electrophoresis of rainbow trout *(RT)* serum diffused against *A. salmonicida* ECP and a trypsin *(Tryp)* solution in the outer troughs

the dissolution of the muscle myosepta extending many cm from the site of injection all closely parallel the description of a furuncle often seen in chronic cases of furunculosis in Atlantic salmon (Mackie and Menzies, 1938).

The effects of ECP on various in vitro fish cell preparations confirm it has acute cytotoxic properties and at higher concentrations is leucocytolytic and haemolytic. However the serum of rainbow trout, brown trout and salmon were found to contain factors which inhibit the haemolytic and proteolytic activity of ECP. An α-migrating globulin of rainbow trout precipitated with trypsin and at least one component of ECP in diffusion experiments and was not C-reactive protein. We believe the inhibition of trypsin and the protease component of ECP by rainbow trout serum plus the results of immunodiffusion is evidence, reported for the first time, that fish, like higher vertebrates (Schultze et al., 1963) contain α-globulins which are protease inhibitors and as such may play an important part in the control of pathophysiological conditions.

The foregoing clearly shows that most, if not all, of the virulence factors of *A. salmonicida* are produced extracellularly. Most strains of *A. salmonicida* produce similar factors but in varying quantity (unpublished results). At present we do not know how many enzymes and components are contained in ECP but it is reasonable to conclude that the protease described by Shieh and MacLean (1975) is one. The amount ($\cong 0.2$ ml) of our ECP preparation required to kill 150 g fish is less than might have been expected when compared with the amounts of serum necessary to neutralise haemolytic activity ($\cong 5:1$) and ECP protease activity ($\cong 7:1$). However the single shot dose may have prevented effective inactivation before irreversible changes occurred or, alternatively, some additional toxic components may be present in the ECP. Additionally, as some proteases are potent inactivators of human plasma α-protease inhibitors (Morihara et al., 1979; Baumstark, 1970) no final conclusions should be reached until the components and their activities of both ECP and fish α-globulins have been characterised.

The numerous reports of furunculosis trial vaccines have concentrated on whole cells, cell wall preparations or protein fractions from cells and in turn success has been measured by the production of agglutinating antibodies and survival of immunized fish. We believe that the apparent failure of such immunogen preparations to protect has resulted from the failure to identify virulence factors and incorporate suitably inactivated forms into such vaccines. The results we report here suggest a new avenue of research for furunculosis vaccine testing.

Acknowledgement. The authors wish to acknowledge the technical assistance of Dominique Gallet de St. Aurin.

References

Anderson DP (1973) Investigations of the lipopolysaccharide fractions from *Aeromonas salmonicida* smooth and rough forms. In: Symposium on the major communicable fish diseases in Europe and their control. FAO, Rome, pp 175–179
Baldo BA, Fletcher TC (1973) C-reactive protein-like precipitins in plaice. Nature (London) 246: 145–146
Baumstark JS (1970) Studies on the elastase-serum protein interaction II. On the digestion of human α_2-macroglobulin, an elastase inhibitor, by elastase. Biochem Biophys Acta 207:318–330
Bøyum A (1968) Separation of leukocytes from blood and bone marrow. Scand J Clin Lab Invest 21: Suppl 97:77
Cruickshank R, Duguid JP, Marmion BP, Swain RHA (1975) In: Medical microbiology, vol II. The practice of medical microbiology. Publ Churchill, Livingstone
Ellis AE, Hastings TS, Munro ALS (1981) The role of *Aeromonas salmonicida* extracellular products in the pathology of furunculosis. J Fish Dis (in press)
Ferguson HW, McCarthy DH (1978) Histopathology of furunculosis in brown trout (*Salmo trutta* L.) J Fish Dis 1:165–174
Fuller DW, Pilcher KS, Fryer JL (1977) A leukocytolytic factor isolated from cultures of *Aeromonas salmonicida*. J Fish Res Board Can 34:1118–1125
Holder IA, Haidaris CG (1979) Experimental studies of the pathogenesis of infections due to *Pseudomonas aeruginosa* extracellular protease and elastase as *in vivo* virulence factors. Can J Microbiol 25:593–599
Karlsson KA (1962) Studies of the haemolysin of *Aeromonas salmonicida*. Nord Veterinaermed 14 Suppl 2
Klontz GW, Yasutake WT, Ross AJ (1966) Bacterial diseases in the Salmonidae in the western United States: Pathogenesis of furunculosis in rainbow trout. Am J Vet Res 27:1455–1460
Liu PV (1957) Survey of haemolysin production among species of Pseudomonads. J Bacteriol 74:718–727
Lowry OH, Rosebrough NJ, Farr AL, Randall RJ (1951) Protein measurement with the Folin phenol reagent. J Biol Chem 197:265–275

Mackie TJ, Menzies WJM (1938) Investigations in Great Britain of furunculosis of the Salmonidae. J Comp Pathol Ther 51:225

McDonald CE, Chen KL (1965) The Lowry modification of the Folin reagent for determination of proteinase activity. Anal Biochem 10:175–177

Morihara K, Tsuzuki H, Oda K (1979) Protease and elastase of *Pseudomonas aeruginosa* inactivation of human plasma a_1-proteinase inhibitors. Infect Immun 24:188–193

Paterson WD (1972) The antibody response of juvenile coho salmon *(Oncorhynchus kisutch)* to *Aeromonas salmonicida* the causative agent of furunculosis. Doct Thesis, Corvallis, Oregon State University

Ross AD (1966) Endotoxin studies. In: Progress in sport fishery research. Res Publ, pp 39–77

Schultze HE, Heismburger N, Heide K, Haupt H, Stariko K, Schwick HG (1963) Preparation and characterisation of a_1-trypsin inhibitor and a_2-plasmin inhibitor of human serum. Proc 9th Congr Eur Soc Haematol. Karger, Basel, pp 1315–1320

Shieh HS, McLean JR (1975) Purification and properties of an extracellular protease of *Aeromonas salmonicida*, the causative agent of furunculosis. Int J Biochem 6:653–656

Wadstrom T, Ljungh A, Wretlind B (1976) Enterotoxin, haemolysin and cytotoxic protein in *Aeromonas hydrophila* from human infections. Acta Pathol Microbiol Scand Sect B 84:112–114

Studies on Vaccination of Atlantic Salmon Against Furunculosis

R. PALMER and P.R. SMITH[1]

Introduction

Furunculosis is endemic in Ireland and, with recent increased investment in salmonid culture, research is being conducted into the applicability and development of furunculosis vaccines and vaccine administration techniques.

A number of studies have been conducted on vaccination against furunculosis using oral administration (reviewed by Corbel, 1975) and inoculation, usually intraperitoneal (IP), with or without adjuvants (Krantz et al., 1963, 1964; Overholser, 1968; Paterson and Fryer, 1974 a,b; Udey and Fryer, 1978). Laboratory and field trials have had some success but, due to the variability of results and difficulties in production and administration, they have not led to a commercially usable vaccine. IP inoculation has, however, been found to be an applicable method of large-scale administration of vaccines (Novotny, 1974; Antipa, 1976; Antipa and Amend 1977). New delivery methods for fish vaccines have been developed in recent years. These include: hyperosmotic infiltration (HI) (Amend and Fender, 1976; Antipa and Amend, 1977; Croy and Amend, 1977), simple immersion of fish into vaccine baths (Egidius and Andersen, 1979; Gould et al., 1979) and spray vaccination (Gould et al., 1978). These techniques have been fully developed for vibriosis vaccination but present published work on their use with furunculosis (Antipa and Amend, 1977) has been inconclusive.

This report describes a field trial of vaccination against furunculosis by IP inoculation, with adjuvant, and HI. The trial used a natural challenge, based on large-scale mortalities experienced during sea-cage rearing of Atlantic salmon, *Salmo salar* (L), in an Irish commercial operation. These mortalities occurred mostly during the first month after transfer of smolts to sea water, and gave a loss of approximately 50% of 30,000 smolts, in each of 2 years of operation. In a third year, the loss of approximately 40% of 15,000 smolts, within 3 weeks of transfer, was attributed predominantly to the release of latent furunculosis following transfer stress (Drinan et al., 1978a). The smolts were supplied by the Electricity Supply Board's (ESB) Parteen Hatchery, which has yearly furunculosis epizootics. Antibiotic treatment for furunculosis at the hatchery and sea farm had proved ineffective due to the appearance of *Aeromonas salmonicida* strains resistant to streptomycin, oxytetracycline and tribressin (potentiated sulphonamide: Wellcome) (Drinan et al., 1978b).

---

1 Department of Microbiology, University College, Galway, Ireland

Materials and Methods

Experimental Fish

Atlantic salmon, yearling pre-smolts (mean weight 40 g) from a common stock, were supplied and held by the ESB Parteen Hatchery.

Vaccine Preparation

Vaccines were prepared from *A. salmonicida* (strain A47R) isolated from a furunculosis outbreak at Parteen Hatchery. Agglutination tests of *A. salmonicida* isolates from this hatchery could demonstrate only one serotype. The strain was an aggregating form (Udey and Fryer, 1978).

Vaccine and test antigens were prepared by culture of the organism in brain heart infusion (BHI: Oxoid) with rotary incubation at 22 °C for 48 h. Samples were cultured on tryptone soya agar (TSA: Oxoid) to check for purity. The broth cultures were killed by addition of formalin (0.6% final concentration) with further incubation at 22 °C for 24 h. Sterility was checked by inoculation of BHI and TSA with culture samples and incubation at 22 °C and 37 °C.

For HI vaccination, a killed broth was diluted 1/3 with sterile, 7.2 pH-buffered hyperosmotic solution, to give a final concentration of: phosphate buffer, 0.2 M, and NaCl, 5.3%. The cell concentration (counting-chamber determination) of the HI suspension was 5.3×10^8 /ml. For IP vaccination, the cells of a killed broth were harvested in a Sharples continuous-flow centrifuge, washed twice with sterile 0.85% saline and resuspended in 0.85% saline to a cell concentration of 8.0×10^{10} /ml. The vaccine was prepared with Freund's complete adjuvant (FCA: Gibco) in a multiple emulsion (Herbert, 1965); a simple emulsion of cell suspension and FCA (1:1) was emulsified to a second phase with an equal volume of sterile 2% Tween 80 (BDH) in 0.85% saline. The emulsion was checked microscopically for uniformity. This preparation gave a free-flowing liquid. Sterile vaccines were stored at 4 °C before use.

Vaccination Procedure

One thousand, five hundred (1,500) fish were vaccinated by HI. Each 7 liters of HI vaccine was used to vaccinate 13.7 kg of fish in 6 lots. Fish were dipped for 1.5 min, the vaccine being continually aerated. One thousand, five hundred (1,500) fish were anaesthetized with M.S.222 (Sandoz), hand-held, and injected IP with 0.1 ml of vaccine with adjuvant (2.0×10^9 cells) using automatic pipetting syringes (1-ml: Becton and Dickinson) with 25 G needles. A group of 1,500 fish were left untreated. Each group was held separately in hatchery tanks, and in cages after sea water transfer. HI and IP vaccination were separated by 2 weeks but all groups were sea-water-transferred simultaneously, 10 weeks post HI vaccination and 8 weeks post IP vaccination.

Diagnosis

Moribund and dead fish were recovered and examined every 2 days. Kidney material was streaked on TSA and TSA with 1% NaCl, and cultured at 22 °C. Isolates were identified by standard biochemical tests.

Antibody Titers

Fish blood samples were taken by syringe from the caudal artery. This was allowed to clot at room temperature, the serum was removed, centrifuged and stored at −20 °C. Agglutinin antibody titers were determined by a microtiter technique using 0.050 ml serum dilutions with equal volumes of a twice-washed, killed cell suspension (0.85 OD at 525 nm) in phosphate-buffered saline (PBS). Microtiter plates were incubated for 4 h at 22 °C followed by overnight at 4 °C.

Results

IP vaccination of 1,500 fish took 2 h, total time, with 3 inoculators; HI vaccination took 40 min, using 2 baths. All fish recovered to normal behaviour within 5 min of being returned to their tanks. IP fish, examined internally for injection lesions, showed, for up to 4 weeks post vaccination, slight inflammation at the site of inoculation and traces of vaccine in the peritoneal cavity. This vaccine/adjuvant was found by microscopic examination to have remained as a multiple-phase emulsion.

No mortalities occurred between vaccination and sea water transfer. Mortalities during a 7-week period after transfer were low in all groups (Table 1), reaching only 107 in total. However, the differences in the total mortalities and the estimated mortalities due to furunculosis between the control and HI groups were significant (chi-squared test; $P<0.001$). No significant (chi-squared test; $\alpha=0.01$) differences in mortalities occurred between the control and IP groups. Other mortalities were attributed to vibriosis or, possibly, transfer stress.

Serum agglutinin antibody titers and water temperatures are presented in Fig. 1. Low levels of antibody occurred in the control fish. Titers in the control and HI groups fell over the period monitored, HI titers were usually higher than control titers

Table 1. Mortalities in vaccinated and control fish over 7 weeks following sea water transfer

Vaccine treatment	No of fish	Mortalities		Estimated mortalities due to furunculosis	
		No	(%)	No	(%)
C	1,500	54	3.9	38	2.5
IP	1,500	41	2.9	19	1.3
HI	1,500	12	0.9	6	0.5

C, control; IP, intraperitoneal injection; HI, hyperosmotic infiltration

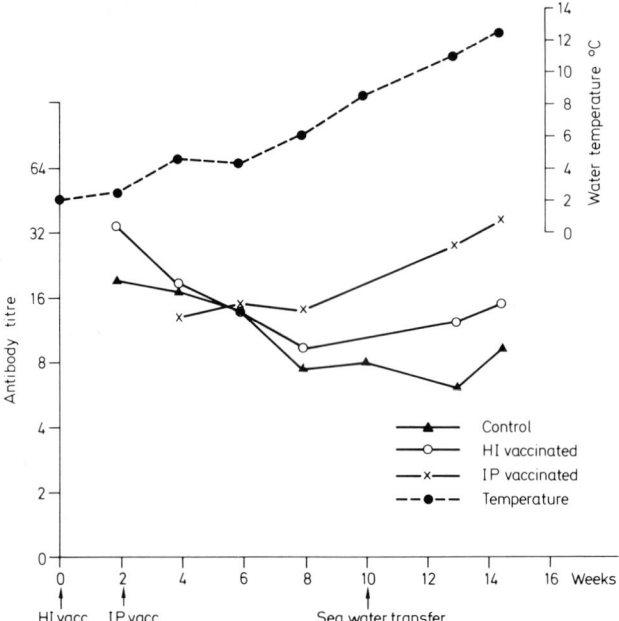

Fig. 1. Serum agglutinin titers against *Aeromonas salmonicida* of vaccinated and control Atlantic salmon. Each point represents the geometric mean of a 10-fish sample (*IP* intraperitoneal injection; *HI* hyperosmotic infiltration)

but this was not significant (F- and t-tests; $\alpha = 0.01$). The IP fish showed a significant (F- and t-tests; $P < 0.005$) rise in titers above the control titers, at 11 weeks post vaccination. This rise showed a possible correlation to increasing water temperature.

Discussion

A significant degree of protection against furunculosis was obtained in Atlantic salmon with HI vaccination but not with IP administration of vaccine with adjuvant. However, the number of mortalities in the control group was very low and, from this data, it is not possible to assess the protection afforded by either technique in a high-challenge situation. The low challenge experienced could possibly be explained by the very low winter and spring temperatures, a more accurate assessment of smoltification time (giving a transfer date of one month earlier than in previous years) and a generally healthier stock (in contrast to previous years, there was a complete lack of fungal infection of fins). It is possible that the level of furunculosis latency was also lower than previously. Levels of latency are being monitored using the prednisalone acetate test of McCarthy (1978). The temperatures at vaccination in this study were lower than anticipated but with stock which are carriers a compromise would be required between higher temperatures, favouring the immune response (Paterson and Fryer, 1974a; Corbel, 1975), and lower temperatures, minimizing the possibility of precipitation of disease. The effects of furunculosis latency on immune responsiveness to furunculosis (or other) vaccines are unknown.

Although using largely unskilled workers, both vaccination techniques were found to be free from difficulties. It is believed that, with experience, labour usage would greatly improve. Antipa (1976) reported that skilled workers could inoculate 600–1,000 22-g fish/h. The lack of any mortalities due to vaccination was encouraging.

No relationship was found between serum agglutinin titers to *A salmonicida* and protection against challenge. There is increasing evidence that fish serum agglutinin titers are not necessarily a reliable index of immunity (Corbel, 1975), Agglutinin titers in HI-vaccinated fish did not show a significant increase. Earlier reports on HI immunization against vibriosis (Antipa and Amend, 1977; Croy and Amend, 1977) gave conflicting findings on agglutinin levels. Krantz et al. (1963, 1964) and Paterson and Fryer (1974b) found increased immune responsiveness to inoculated *A. salmonicida* antigens with incorporation of the antigen into FCA. The use, in this study, of a multiple-phase emulsion of the vaccine with FCA overcame practical problems associated with the viscosity of simple water-in-oil emulsions. The stability and slow clearance of this emulsion was shown by its persistence over several weeks in the peritoneal cavity of inoculated fish. The delay between inoculation and increase in antibody levels may be a function of this slow release (allowing formation of antigen-antibody complexes; Paterson and Fryer, 1974b) and/or the low temperatures at inoculation. Natural agglutinins to *A. salmonicida* occurred in the control group; the specificity of these was not determined, however. Krantz and Heist (1970) found a positive correlation of natural antibody with previous exposure to the pathogen.

In a current trial the role of the hyperosmotic agent and broth culture components in furunculosis immunization are being investigated. Groups of 300 Atlantic salmon fry were immersion-vaccinated with (buffered, formalinized) whole culture, culture supernatant or resuspended washed cells; with or without incorporated hyperosmotic agent. Sham-vaccinated control groups were included. Protection will be assessed by laboratory challenge.

Acknowledgements. We wish to thank Noel Roycroft of the Electricity Supply Board and Padraig de Bhaldraithe of Bradan Mara Teo. for their practical help and encouragement. We also thank Bill Sheeran, John O'Kelly, Evelyn Drinan, and Seamus Dunne of the Fish Disease Laboratories, University College, Galway.

References

Amend DF, Fender DC (1976) Uptake of bovine serum albumin by rainbow trout from hyperosmotic solutions: a model for vaccinating fish. Science 192:793–794

Antipa R (1976) Field testing of injected *Vibrio anguillarum* bacterins in pen-reared Pacific salmon. J Fish Res Board Can 33:1291–1296

Antipa R, Amend DF (1977) Immunization of Pacific salmon: comparison of intraperitoneal injection and hyperosmotic infiltration of *Vibrio anguillarum* and *Aeromonas salmonicida* bacterins. J Fish Res Board Can 34:203–208

Corbel MJ (1975) The immune response in fish: a review. J Fish Biol 7:539–563

Croy TR, Amend DF (1977) Immunization of sockeye salmon *(Oncorhynchus nerka)* against vibriosis using the hyperosmotic infiltration technique. Aquaculture 12:317–325

Drinan E, Palmer R, O'Kelly J, Smith PR (1978a) Investigation of mortalities of salmon smolts on their transfer to the sea. Fish Pathology Group Report I University College, Galway, pp 13

Drinan E, Sheeran W, Smith PR (1978b) Antibiotic resistance of strains of *Aeromonas salmonicida* isolated from hatchery-reared salmon. Fish Pathology Group Report II University College, Galway, p 5

Egidius EC, Andersen K (1979) Bath-immunization — a practical and non-stressing method of vaccinating sea-farmed rainbow trout *Salmo gairdneri* Richardson against vibriosis. J Fish Dis 2:405–410

Gould RW, O'Leary PJ, Garrison RL, Rohovec JS, Fryer JL (1978) Spray vaccination: a method for the immunization of fish. Fish Pathol 13:63–68

Gould RW, Antipa R, Amend DF (1979) Immersion vaccination of sockeye salmon *(Oncorhynchus nerka)* with two pathogenic strains of *Vibrio anguillarum*. J Fish Res Board Can 36:222–225

Herbert WJ (1965) Multiple emulsions: a new form of mineral-oil antigen adjuvant. Lancet 2:771

Krantz GE, Heist CE (1970) Prevalence of naturally acquired agglutinating antibodies against *Aeromonas salmonicida* in hatchery trout in central Pennsylvania. J Fish Res Board Can 27:5

Krantz GE, Reddecliff JM, Heist CE (1963) Development of antibodies against *Aeromonas salmonicida* in trout. J Immunol 91:757–760

Krantz GE, Reddecliff JM, Heist CE (1964) Immune response of trout to *Aeromonas Salmonicida*. I. Development of agglutinating antibodies and protective immunity. Prog Fish Cult 25:3–10

McCarthy DH (1978) Some ecological aspects of the bacterial fish pathogen *Aeromonas salmonicida*. In: Skinner FA, Shewan JM (eds) Aquatic microbiology. Academic Press, London New York, pp 299–324

Novotny AJ (1974) Salmon vaccinated to prevent vibriosis. NW Fisheries Center Monthly Report, US Department of Commerce NOAA and NMFS, Seattle, pp 1–5

Overholser DL (1968) Control of furunculosis in Pacific salmon by immunization. MSc Thesis, Oregon State University, p 58

Paterson WD, Fryer JL (1974a) Effect of temperature and antigen dose on the antibody response of juvenile salmon *(Oncorhynchus kisutch)* to *Aeromonas salmonicida* endotoxin. J Fish Res Board Can 31:1743–1749

Paterson WD, Fryer JL (1974b) Immune response of juvenile coho salmon *(Oncorhynchus kisutch)* to *Aeromonas salmonicida* cells administered intraperitoneally in Freund's complete adjuvant. J Fish Res Board Can 31:1751–1755

Udey LR, Fryer JL (1978) Immunization of fish with bacterins of *Aeromonas salmonicida*. Mar Fish Rev 40(3):12–17

Further Studies on Furunculosis Vaccination

P.D. SMITH[1], D.H. McCARTHY[2], and W.D. PATERSON[3]

Introduction

The need for an economically viable and easily administered vaccine against infectious bacterial diseases in farmed fish has been recognised for many years. The majority of these diseases have been contained by the use of bacteriostatic or bacteriocidal agents; these are costly and of limited use when a disease has become firmly established.

Oral vaccines have been developed for the treatment of fish diseases, with varying degrees of success. Duff (1942) was the first to demonstrate induction of resistance to furunculosis by oral vaccination. However, later work by Snieszko and Friddle (1949), Post (1963), and Krantz et al. (1964) failed to produce resistance to furunculosis in fish given oral vaccines containing heat- or chloroform-killed organisms. Ross and Klontz (1965) and Fryer et al. (1976) have demonstrated protection in fish against Hagerman redmouth disease and furunculosis respectively.

A new technique for the immunization of fish was put forward by Amend and Fender (1976). They demonstrated that bovine serum albumin could be made to infiltrate the blood system of rainbow trout *(Salmo gairdneri)* by a method called hyperosmotic infiltration (HI), and they concluded that this technique was a possible model for fish vaccination. As a result of this work Antipa and Amend (1977) made a comparison of the protection afforded by intraperitoneal-injected and HI vaccines against *Vibrio anguillarum* and *Aeromonas salmonicida* antigen. Their results showed a degree of protection against vibriosis with both vaccination methods tested, the HI method of vaccination being more successful. Unfortunately, no results for efficacy of the protection against furunculosis were obtained as no significant epizootic occurred.

Conflicting evidence is also available as to the type of immune response elicited by oral and HI vaccines. While most studies have involved analysis of the humoral immune response, no information is available on the cellular immune response. Thus Duff (1942) found an increase in circulating antibodies in fish given an oral vaccine, while

---

1 Department of Bacteriology and Virology, University of Manchester, Oxford Road, Manchester M13 9PT, England
2 Ministry of Agriculture, Fisheries and Food, Directorate of Fisheries Research, Fish Diseases Laboratory, Weymouth, Dorset DT4 8UB, England,,
  Present address: Tavolek Laboratories Inc., 2779 152nd Ave, N.E., Redmond, Washington 98052, USA
3 Environment Canada, Fisheries and Marine, Halifax Laboratory, Nova Scotia, Canada B3J 2R3
  Present address: Connaught Laboratories Ltd, 1755 Steeles Avenue West, Willowdale, Ontario M2N 5T8, Canada

Krantz et al. (1964) failed to stimulate detectable serum antibody levels in fish treated with a similar vaccine. In another group of experiments, Post (1963) found that although fish given an oral vaccine against *Aeromonas hydrophila* had agglutinating antibody in their serum, upon challenge these fish had negligible protection. On the other hand, Ross and Klontz (1965) were able to protect fish by oral immunization, but were unable to detect antibody in immunized fish.

The type of immune response stimulated by HI vaccination is also in doubt. In the experiments of Antipa and Amend (1977), serum agglutinin titres were similar to those produced by injection. However, these workers questioned whether serum agglutinin levels are a reliable index of the degree of disease immunity.

This paper describes one oral vaccine against furunculosis and two types of vaccine introduced by hyperosmotic infiltration, the protection afforded by them during a spontaneous epizootic in "natural" conditions, and the nature of the immune response stimulated by such vaccines.

Materials and Methods

Vaccine Preparation

Thre vaccines were prepared:
1) An oral vaccine consisting of whole formalized cells of *A. salmonicida* (isolate A) incorporated into fry food
2) A vaccine consisting of whole bacterial cells of *A. salmonicida* (isolate B) disrupted by treatment with sodium dodecylsulphate
3) A vaccine consisting of whole bacterial cells of *A. salmonicida* (isolate B) disrupted by ultrasonication

Vaccine 1 was administered to brown trout *(Salmo trutta)* fry over a period of 30 days (McCarthy, Paterson and Smith, to be published). Vaccines 2 and 3 were administered to brown trout fry by the hyperosmotic method of vaccination (McCarthy et al., to be published). Samples of all three vaccine preparations were tested for sterility in the laboratory before field use.

Field Trials

After a period of observation, the three groups of vaccinated fish and their controls were transferred to a fish farm chosen for its history of regular annual outbreaks of furunculosis. Fish mortalities in the various groups were monitored throughout the summer.

Tests of Immunity

An analysis of three possible types of immune response to the vaccines was made as follows:

1. Circulating antibody. Serum antibody was determined either by standard microtitre procedures using formalin-killed *A. salmonicida* or by a modification of

Hansen and Lingg's (1976) latex agglutination method for the detection of antibodies against *A. salmonicida* antigens. Briefly, the latter involved the coating of latex particles with a boiled aqueous extract of *A. salmonicida*. The coated particles were then incubated with doubling dilutions of antiserum in microtitre trays and the reactions read after 15 min. Non-sensitized latex particles and normal trout sera were used as controls for non-specific agglutination. The latex agglutination method provided an extremely sensitive test for detection of antibody levels.

2. Secretory antibody. Mucus was removed from the gut of vaccinated and control fish by vigorous washing with balanced salt solution. Samples were then analysed for the presence of secretory antibodies against *A. salmonicida* using the latex agglutination test.

3. Cell-mediated immunity. Leucocytes from vaccine-treated and control fish were harvested from heparinized peripheral blood using Ficoll-Paque (Pharmacia Ltd.) and the cellular immune response to *A. salmonicida* was measured using a modification of Søberg and Bendixen's (1967) leucocyte migration inhibition test (Smith, to be published). This involved the packing of leucocytes into capillary tubes and the incubation of these tubes in culture chambers in the presence or absence of the antigen under examination. Under normal conditions (i.e. in the absence of specific sensitizing antigen), the leucocytes migrate out of the capillary tubes in the form of a fan. The extent of migration of leucocytes from the capillary tubes served as an indicator of the reaction between antigen and specifically sensitized lymphocytes present in the leucocyte suspension. Thus, if antigen to which lymphocytes were sensitized was present in the incubation medium, then such lymphocytes produced leucocyte migration inhibition factor (LMIF) which inhibited the normal emigration of leucocytes from the capillary tubes in which they were packed.

Results

Comparison of Mortalities Between Oral Vaccine Hyperosmotic Vaccines and Control Groups

In a natural epizootic fish treated with the oral vaccine showed a 35% cumulative mortality, the SDS-vaccinated group 62% mortality, the USD-vaccinated group 62% mortality, and the control fish experienced an 86% loss (McCarthy et al., to be published).

A bacteriological study was made of the fish which had succumbed to furunculosis in these trials and a strain of *A. salmonicida* (isolate C), distinct from the two strains used in the production of the three vaccines, was isolated.

Circulating Antibody Levels

Prior to challenge, serum samples from fish of each vaccinated group were tested for the presence of agglutinating antibody. Table 1 shows that fish from the three vaccinated groups did not have levels of circulating antibodies significantly different from those of control fish.

Table 1. Serum antibody titres for vaccine-treated and control fish

Treatment	No of fish	No of specimens with indicated endpoint dilutions							
		$<\frac{1}{10}$	$\frac{1}{10}$	$\frac{1}{20}$	$\frac{1}{40}$	$\frac{1}{80}$	$\frac{1}{160}$	$\frac{1}{320}$	$\frac{1}{640}$
Oral vaccine	25	21	2	0	2	0	0	0	0
Oral control	30	28	2	0	0	0	0	0	0
USD HI vaccine	20	15	3	2	0	0	0	0	0
SDS HI vaccine	20	18	0	1	1	0	0	0	0
HI control	28	20	5	3	0	0	0	0	0

Secretory Antibody

Mucus from gut washings was taken from fish in vaccinated and control groups and examined for secretory antibody using the latex agglutination test. In no case was a significantly high titre of antibody against *A. salmonicida* found in gut mucus.

Cell-Mediated Immunity

Leucocyte migration inhibition values in the presence of either of the vaccine strains (A or B) or the natural-challenge strains (C) for leucocytes taken from fish which had received the oral vaccine, the two types of hyperosmotic vaccine, and the vaccine controls, respectively, are shown in Table 2. Numbers of fish showing positive leucocyte migration inhibition (>30% inhibition) against the vaccine strain and natural-challenge strain are given. Results are also given for leucocyte migration inhibition values in the

Table 2. Leucocyte migration inhibition results for vaccinated and control fish

Treatment	Number of fish showing positive leukocyte migration inhibition in presence of antigen (% in brackets)				
	Isolate A	Isolate C	Isolate B	Pseudomonas	PHA
Oral vaccine	38/36 (77%)	20/36 (55%)	–	4/31 (13%)	29/29 (100%)
Oral vaccine control	3/30 (10%)	1/30 (3%)	–	2/26 (8%)	23/23 (100%)
USD HI vaccine	–	10/25 (40%)	18/25 (72%)	1/20 (5%)	9/9 (100%)
SDS HI vaccine	–	8/24 (33%)	19/25 (76%)	2/20 (10%)	12/12 (100%)
HI control	–	2/26 (8%)	1/25 (4%)	1/20 (5%)	20/20 (100%)

presence of a strain of *Pseudomonas fluorescens* and in the presence of phytohaemagglutinin (PHA).

It is of interest that, while a high percentage of fish showed leucocyte migration inhibition in the presence of the bacterial strain incorporated into the vaccine, a much lower percentage showed a cellular response against the natural-challenge strain. It should also be noted that, although numbers are small, the percentage of fish producing a positive cellular response against the natural-challenge strain corresponded very closely to the percentage found to be resistant to furunculosis in the vaccine field trials. In all cases, PHA (a substance which stimulates lymphocytes to produce LMIF) produced marked inhibition of leucocyte migration.

Leucocyte migration inhibition tests were repeated on vaccine-treated fish kept in *A. salmonicida*-free water under laboratory conditions 6, 12, and 18 months after vaccination and, in all cases, it was noted that the percentage of fish showing a cell-mediated immune response against bacterial strains decreased with time after vaccination. Indeed, at 18 months after vaccination, leucocyte migration inhibition in the presence of bacterial antigens could only be detected in a very small number of fish. It was not possible to determine whether this decrease in the cellular immune response was reflected by a decrease in resistance to furunculosis.

Discussion

All three vaccines showed some degree of success, the oral vaccine giving a greater degree of protection than the two hyperosmotically introduced vaccines. A more detailed analysis of the mechanisms involved in the uptake of antigen by the hyperosmotic method of vaccination using radiolabelled substances in now being undertaken at the Weymouth laboratory. Results obtained suggest that the main portal of entry of vaccine is the gill tissue and that particulate antigens are more rapidly taken up from the vaccine bath than soluble antigens. It is hoped that further study of the HI technique could lead to an improvement of this method.

Although both oral and hyperosmotic vaccines have been used by previous workers, there are still conflicting opinions as to the immune mechanisms involved in protection. While circulating antibody has been found in some experiments with oral vaccination, other studies have noted an absence of such antibody. Indeed, Fryer et al. (1978), working on an oral vaccine for vibriosis, suggested that lack of circulating antibody may reflect the fact that other mechanisms such as secretory antibody or cellular immunity may be of importance. Similarly, there is some doubt as to the immune mechanism elicited by hyperosmotic vaccination. Antipa and Amend (1977) found increased levels of *V. anguillarum* serum agglutinins in salmon after treatment with a vibrio hyperosmotic vaccine. However, they questioned whether serum agglutinin levels were a reliable index of the degree of disease immunity.

In this study, although a resistance to furunculosis was produced by all three vaccines, a corresponding increase in either serum or secretory antibody was not found, suggesting that these mechanisms do not play a major role in conferring immunity to vaccinated fish. When the cellular immune response was studied using a modified leucocyte migration inhibition test, however, a close association between disease resistance and cell-mediated immunity was found. While further investigations must be made to confirm whether cellular immunity does confer resistance to furunculosis in vaccinated fish, application of the leucocyte migration inhibition test may well be a good measure of the efficacy of such vaccines.

It is of interest that while the oral vaccine appeared to stimulate a cell-mediated immune response, no secretory antibody was produced. Fletcher and White (1973) reported the presence of secretory antibody in the intestines of plaice, *Pleuronectes platessa* L., after oral administration of *V. anguillarum* antigens. No such secretory antibody was found in gut mucus of fish receiving oral vaccine in the present experiments, even when examined by such a sensitive test as the latex agglutination technique.

The findings that immunization with a vaccine containing one strain of *A. salmonicida* does not necessarily confer an overall protection to a different strain encountered in a natural-challenge situation have considerable implications in the development of vaccine against furunculosis. Results reported here suggest that an improved oral or hyperosmotic vaccine should be of a multivalent type containing several representative strains of *A. salmonicida*. Alternatively, work in progress at the Weymouth laboratory has suggested that while there may be diversity between strains of *A. salmonicida*, which could affect their immunogenic characteristics, all strains produce an extracellular product (ECP) which shows little antigenic variability. This toxic cell-free supernatant from *A. salmonicida* may be a mixture of ECP and perhaps other surface-derived antigens such as lipopolysaccharides (LPS). Antiserum raised against this supernatant produced by one strain of *A. salmonicida* will cross-react strongly with ECPs produced by other strains of the organism. Furthermore, intramuscular injection of such products into fish will produce a lesion macroscopically and microscopically similar to that seen in natural furunculosis, the fish developing antibodies which cross-react with ECP.

It has proved possible to inactivate the toxic properties of the ECP by treatment with formaldehyde solution and, while this "inactivated ECP" does not produce furunculosis-like lesions on injection into fish, it does stimulate production of antibody which will cross-react with "active ECP". A vaccine incorporating this inactivated ECP is now under development.

References

Amend DF, Fender D (1976) Uptake of bovine serum albumin by rainbow trout from hyperosmotic solutions: a model for vaccinating fish. Science 192:793–794

Antipa R, Amend DF (1977) Immunization of Pacific Salmon: Comparison of intraperitoneal injection and hyperosmotic infiltration of *Vibrio anguillarum* and *Aeromonas salmonicida* bacterins. J Fish Res Board Can 34:203–208

Duff DCB (1942) The oral immunization of trout against *Bacterium salmonicida*. J Immunol 44:87–93

Fletcher TC, White A (1973) Antibody production in plaice (*Pleuronectes platessa* L.) after oral and parenteral immunization with *Vibrio anguillarum*. Aquaculture 1:417–428

Fryer JL, Rohovec JS, Tabbit GL, McMichael JS, Pilcher KS (1976) Vaccination for control of infectious diseases in Pacific salmon. Fish Pathol 10:155–164

Fryer JL, Rohovec JS, Garrison RL (1978) Immunization of salmonids for control of vibriosis. Mar Fish Rev 40(3):20–23

Hansen CB, Lingg AJ (1976) Inert particle agglutination test for detection of antibody to enteric redmouth bacterium. J Fish Res Board Can 33:2857–2860

Krantz GE, Reddecliffe JM, Heist CE (1964) The immune response of trout to *Aeromonas salmonicida*. II Evaluation of feeding techniques. Prog Fish Cult 26:65–69

Post G (1963) The immune response of rainbow trout *(Salmo gairdneri)* to *Aeromonas hydrophila*. In: Dep Inf Bull Utah, State Dep Fish Game 64:7

Ross AJ, Klontz GW (1965) Oral immunization of rainbow trout *(Salmo gairdneri)* against an etiologic agent of "redmouth disease". J Fish Res Board Can 22:713–719

Smith PD (1980) A modified leucocyte migration inhibition test for the detection of cellular responses in brown trout *(Salmo trutta)* to be published

Snieszko SF, Friddle SB (1949) Prophylaxis of furunculosis in brook trout *(Salvelinus fontinalis)* by oral immunization and sulphamerazine. Prog Fish Cult 113:161–168

Søberg M, Bendixen G (1967) Human lymphocyte migration as a parameter of hypersensitivity. Acta Med Scand 181:247–253

Pathogenesis of Carp Erythrodermatitis (CE): Role of Bacterial Endo- and Exotoxin

J.M.A. POL, R. BOOTSMA, and J.M. v. d. BERG-BLOMMAERT[1]

Introduction

Carp erythrodermatitis (CE) is a subacute to chronic contagious disease of the skin. The infection frequently starts at the site of an injury to the epidermis. A hemorrhagic inflammatory process then develops between the epidermis and the dermis. As the infection spreads the red inflammatory zone gradually extends; tissue breakdown leads to the formation of a central ulcer. In terminal stages a generalized edema may occur.

Fijan (1972) suggested separation of the disease as a distinct pathological entity from the infectious dropsy of carp (IDC) complex, and gave it the present name.

The etiology of CE was established by Bootsma et al. (1977). The causative bacterium was preliminarily characterized as a nonmotile *Aeromonas* species. During later experiments [3, 9] it was demonstrated that the CE bacterium should be assigned to the *Aeromonas salmonicida* complex but, using the present taxonomic subdivision [6], it cannot be identified at a subspecies level.

At bacteriological examination of diseased carp it was repeatedly noted that only low numbers of the CE bacterium can be found in the skin lesion (Bootsma and Blommaert, unpublished). This finding was confirmed by examining skin sections using the FAT technique (Vos-Maas, unpublished). In several cases, bacteremia could only be demonstrated during premortal stages of the disease. These findings led to the hypothesis that some toxic factor released by the CE bacterium could account for the marked inflammation, the tissue necrosis and, possibly, for the generalized edemas.

In the present paper the pathologic effects of endotoxin extracted from the CE bacterium, and cell-free culture supernatant were studied. Carp and mice were used as test animals. In addition, a few preliminary tests were carried out in order to characterize the toxic factor in the culture supernatant.

Materials and Methods

Test Animals

Balb/c mice (TNO, Zeist, The Netherlands), 4 weeks old, were housed in macrolon plastic cages. Room temperature was 23 °C; air humidity was maintained at 50% saturation. Pelleted food (Muracon, Trouw Ltd, Putten, The Netherlands) and water

---

[1] Department of Special Animal Pathology, Utrecht, Netherlands

were administered ad libitum. One-summer-old mirror carp (*Cyprinus carpio* L.), weighing 100–200 g each, were kept in 20–200-l aquaria. The aquaria were aerated and continuously supplied with running tap water of 25 °C. The carp were fed a pelleted trout food (Trouvit, Trouw Ltd) 4–5 times a day. All test fish had been kept for several months in aquaria without showing any clinical sign of disease.

The Bacterium

CE isolate V 76/134 was used throughout this study. Cultures were maintained in a semisolid medium at 12 °C, as described previously [2].

Endotoxin

Bacterial cells were grown in 200 petri dishes (10 cm ⌀) on a solid medium composed of tryptose (Difco) 1% and agar 1.5%, supplemented with horse blood serum 10% (v/v). Approximately 30 g of cells (wet weight) was harvested after 3x24 h at 27 °C. Endotoxin (lipopolysaccharide, LPS) was extracted from the cells using the phenol extraction method [8].

Total LPS yield was 140 mg dry weight. Next the LPS was dissolved in pyrogen-free phosphate-buffered saline (PBS). The LPS solution (6 mg/ml) was inoculated intravenously (IV) and intraperitoneally (IP) into mice, and IP into carp.

Cell-free Culture Supernatant

Bacterial cells were grown in liquid media containing 1% tryptose, supplemented with either 0.5% synthetic seasalt or 10% (v/v) horse blood serum. After 3x24 h at 27 °C cells were removed by centrifugation for 30 min at 3,000 G. The supernatant was filtered through a 110-nm Millipore filter. Tests for the presence of bacterial cells were made by incubating samples at 27 °C. Next the supernatant was supplemented with kanamycin 50 μg/ml (MIC=2 μg/ml) and stored at 4 °C. Toxicity tests were made by IP, IV, and subcutaneous (SC) inoculation into mice, and IP, intracardial, and subepidermal inoculation into carp.

Preliminary Characterization of Toxic Factor(s) in the Culture Supernatant

Preliminary information on the molecular size of the toxic factor was obtained by transferring the tryptose-seasalt supernatant to a dialysis bag (cellulose tube, pore size 40 Å, Visking, Zuid Holland Ltd, The Hague, The Netherlands) and concentrating the volume ten times in a vacuum container. To remove excess of salt, the content of the bag was then dialyzed against running tap water. Subsequently, the contents of the dialysis bag and of the vacuum container were separately inoculated IP into carp.

Indications for a protein nature of the toxic factor were obtained by exposing different batches of tryptose-serum supernatant to heat (30 min, 60 °C), a low pH, formalin treatment, and ammonium sulfate precipitation, followed by IP inoculation into carp. Acidification to pH 2.0 was performed by adding 1 N HCl. After 30 min the pH was adjusted to 7.0 with 1 N NaOH, followed by dialysis against running tap water to remove excess of salt. Formalin treatment was carried out by supplementing supernatant with 10% formaldehyde to obtain a final concentration of 2%. After 24 h exposure at 4 °C the formalin was removed by dialysis against running tap water. Ammonium sulfate, saturated solution, was slowly added to supernatant, to obtain a final concentration of 50%. After stirring the precipitate was sedimented by centrifugation for 30 min at 1,700 G. The remaining supernatant and a solution of the precipitate in water were separately dialyzed against running tap water, to remove excess of ammonium sulfate. The supernatant and the dissolved precipitate were inoculated IP into carp.

Clinical and Pathological Examinations

Hematocrit values were determined using heparized glass capillaries. Immediately after filling the capillaries were centrifuged at 15,000 G for 5 min. Protein concentrations of blood plasma and oedematous fluids were determined using an Atago SPRt2 refractometer.

For histological examination tissue samples were fixed in Bouin Hollande and processed using standard techniques. The samples were embedded in Paraplast (Sherwood Med Ind, St. Louis, Missouri, USA). Sections were stained with Mayer's hematoxylin and eosin.

Results

Endotoxin

LPS was toxic to mice. At 20 mg/kg IV the mice showed clinical signs of fever and distress, i.e., a rough hair-coat, loss of appetite, and apathy. After 3 days the symptoms gradually disappeared and the mice recovered. At 40 mg/kg IV the same effects were noted, but they occurred more acutely, followed by death within 24 h. The same dose IP was also lethal.

LPS had no perceptible effect on carp at dosages of 40 and 80 mg/kg IP.

Cell-free Culture Supernatant: Toxicity to Mice

IV inoculation of 0.1 ml tryptose-serum supernatant produced clinical signs of fever, followed by death after 4 days. At autopsy hemorrhages were found in the thoracic and in the abdominal cavity. At 0.1 ml IP the same symptoms were noted, but the mice survived and had fully recovered after 14 days. Autopsy, performed 4 days after inoculation, revealed subperitoneal hemorrhages. SC inoculation of 0.05 ml resulted in a local inflammatory response with vasodilatation and hemorrhages.

Cell-free Culture Supernatant: Toxicity to Carp

Subepidermal injection of 0.01 ml tryptose-serum supernatant produced a local CE-like lesion: subepidermal vasodilatation and hemorrhages, followed by necrosis and ulceration.

Tryptose-serum supernatant was lethal to carp at a dose of 0.5 ml/100 g IP or intracardially. After IP inoculation, some of the fish showed extensive vasodilatation, hemorrhages and edemas. These effects were first observed around the site of injection, suggesting an initial spread of the toxic factor through the abdominal wall. After 2–3 days generalized vasodilatation and edemas, suggestive of a toxemia, were recorded. Some of the fish only showed blackening of the skin. All IP inoculated carp died after a maximum of 29 days.

In case of severe hemorrhages hematocrit values had dropped to 20%, whereas values of 45%–55% were found in control fish. In edematous fish serum protein values had decreased to 1.5–2.5 g%, as compared with 4.5–6.5 g% in control fish. Ascitic fluid contained 1.5–2.5 g% protein; edematous fluids from the eye orbits and scale pockets contained 1.0–1.5 g% protein. Histologically, the findings closely resembled those occurring with natural CE (Frederix-Wolters, unpublished). It is noteworthy that after IP inoculation CE-like lesions were also found subepidermally (Fig. 1).

Pathologic effects produced by tryptose-seasalt supernatant were essentially the same as those produced by tryptose-serum supernatant. The lethal dose was a little

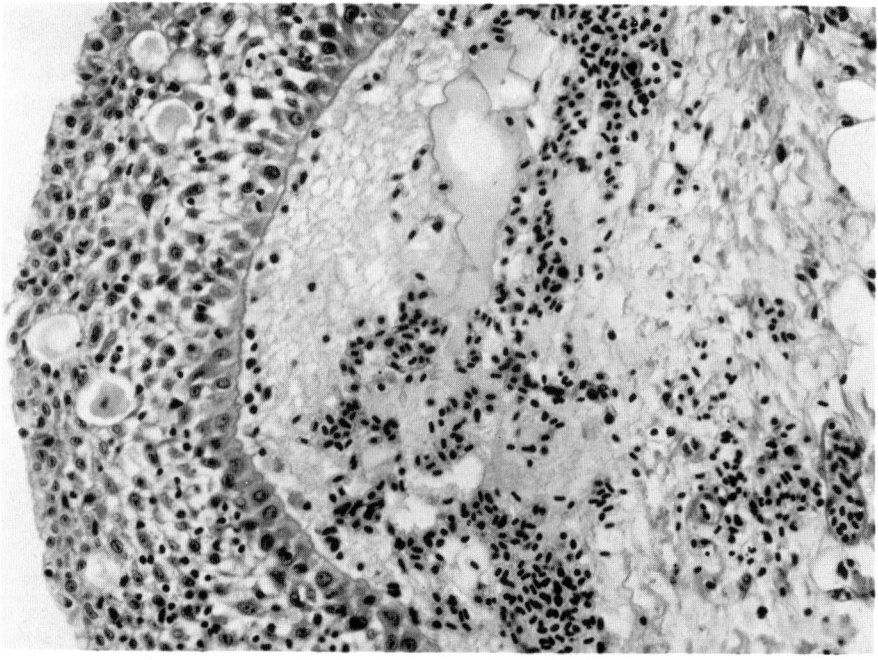

Fig. 1. Histologic section through the skin of a mirror carp injected IP with 0.5 ml culture supernatant, HE stain ×750. Accumulation of blood cells and fluid between epidermis *(left)* and dermis *(right)*

higher, but after IP inoculation generalized vasodilatation and edemas occurred within 24 h, followed by death within 48 h. After supplementing tryptose-seasalt supernatant with 10% (v/v) horse blood serum and incubating the mixture for 24 h at 27 °C, the occurrence of generalized effects was retarded as with tryptose-serum supernatant. This finding indicates that the toxic factor was probably bound to serum proteins.

Preliminary Characterization of the Toxic Factor in the Culture Supernatant

The toxic factor did not pass the wall of the dialysis bag, indicating a MW <15,000 daltons. The tenfold-concentrated tryptose-seasalt supernatant appeared to be lethal at dosages much lower than 0.5 ml/100 g.

The toxic factor was inactivated by heat, a low pH, and formalin treatment. IP inoculation of 1 ml supernatant/100 g, after various treatments, had no perceptible effect on carp. These findings are suggestive of a protein nature.

The toxic factor was removed from tryptose-serum supernatant by ammonium sulfate precipitation. The precipitate was as toxic as the initial supernatant. Possibly the protein-bound toxic factor was precipitated together with serum proteins. Attempts to precipitate the toxic factor in tryptose-seasalt supernatant failed.

Discussion

Endotoxin

The negative results with bacterial LPS are in accordance with those obtained by others [1], who found that *Escherichia coli* LPS is not toxic to carp at 80 mg/kg. Apparently endotoxin is less toxic to carp than it is to certain mammals.

Exotoxin

If a toxic substance from a pathogenic bacterium is to be implicated as a determinant of virulence, it must be demonstrated to produce one or more of the specific symptoms of the disease. Furthermore the site of action and the effective concentration must be such that they could plausibly be obtained in the course of a natural infection [7]. All bacterial exotoxins that have been characterized chemically have been found to be proteins. Exposure of exotoxins to heat, acid, and formaldehyde eliminates toxicity [5]. In view of these criteria, the results obtained with culture supernatant strongly suggest that the lesions occurring with natural CE are partly or exclusively produced by a genuine bacterial exotoxin. The generalized edemas, frequently encountered during premortal stages of natural CE, would then be indicative of toxemia.

Natural and Artificial Infection

It is assumed that natural infection takes place after an injury to the epidermis [4]. In addition, artificial infection should preferably be carried out using the method of skin scarification [2, 4]. In such tests only the bacterium itself is introduced into the skin, the exotoxin being washed away. Inoculation of bacterial cells in liquid culture medium into carp will result in higher mortality and less specific disease symptoms.

Immunization

Because of the important role of exotoxin in the pathogenesis of CE, an antitoxic immunity may be equally or even more protective than immunity against the bacterium itself. This point should be investigated further when considering the preparation of a vaccine.

Acknowledgment. The authors wish to express their sincere thanks to Mr. D.J. Kool, Mr. C. Dekker, and Mr. B. de Graaf for technical assistance.

References

1. Berczi I, Bertok L, Bereznai T (1966) Comparative studies of the toxicology of Escherichia coli LPS endotoxin in various animal species. Can J Microbiol 12:1070–1071
2. Bootsma R, Fijan NN, Blommaert J (1977) Isolation and preliminary identification of the causative agent of Carp Erythrodermatitis. Vet Arhiv 47(6):291–302
3. Bootsma R, Blommaert J (1978) Zur Aetiologie der Erythrodermatitis beim Karpfen Cyprinus carpio L. In: Neuere Erkenntnisse über Fischinfektionen. Gustav Fischer, Stuttgart New York, S 20–27
4. Fijan NN (1972) Infectious dropsy in carp – a disease complex. In: Mawdesley-Thomas LE (ed) Diseases of fish. Academic Press, London New York, pp 39–51
5. McCarty M (1973) Host – parasite relations in bacterial diseases. In: Davis BD, Dulbecco R, Eisen HN, Ginsberg HS, Wood WB (eds) Microbiology. 2nd edn. Harper & Row, Hagerstown, USA, pp 635–638
6. Schubert RHW (1974) Chapter on the genus Aeromonas. In: Buchanan RE, Gibbons NE (eds) Bergey's manual of determinative bacteriology, 8th edn. Williams and Wilkins Company, Baltimore, pp 345–348
7. Stanier RY, Doudoroff M, Adelberg EA (1971) Chapter on bacterial exotoxins. In: General microbiology. MacMillan Press Ltd, London, pp 786–789
8. Westphal O, Lüderitz O, Bister F (1952) Über die Ekstraktion von Bakterien mit Phenol/Wasser. Forschung 7b:148–155
9. Wiedemann H (1979) Erythrodermatitis der Karpfen – zur Isolierung und Klassifizierung des Erregers. Dtsch Tierärztl Wochenschr 86:176–181

Examination of the CE Agent

G. CSABA, B. KÖRMENDY, and L. BEKESI[1]

Abstract

Over the last four years 15 bacterial isolates of CE agent were obtained in our examinations from ulcers of carp from various Hungarian fish farms. Successful isolations were primarily carried out from the periferal area of ulcers on blood agar. Then liquid medium cultures after 72 h of incubation were used for experimental inoculations into the scarified skin of carp using injection needles for this purpose. Infected carp showed characteristic dermatitis 5–6 days following scarification. There were either ulcers, sometimes even in small groups along the line of the scarification, or small-sized papulas with hyperemic margins on the sites of needle-pricking, and the central area of the latter was often necrotic. Similar results were also obtained following experimental infection into the damaged fins of carp, which then often broke and came off. The pathogenic agent was easily reisolated from the lesions. The 15 isolates had almost identical features in the bacteriological procedures. According to our examinations these CE agents seemed to belong to the *Aeromonas* genus of Vibrionaceae. More exactly, their bacteriological properties mainly corresponded to those of *Aeromonas salmonicida* with some small differences from its subspecies.

---

1 Central Veterinary Institute, 1581 Budapest, Pf. 2, Hungary

Some Aspects of the Histopathology of Carp Erythrodermatitis (CE)

E. K. GAYER, L. BEKESI, and G. CSABA[1]

The ulcerous skin-inflammation of carps (CE) occurs in Hungary all year round. Based on data collected from different parts of the country the diagnostic work successfully isolated a uniform species of *Aeromonas salmonicida*. The experimentally produced tissue changes caused by this pathogen were traced to and compared with results found in cases of naturally acquired disease and with parasitic changes of the skin. In order to follow up the healing process and to study the regeneration, enriched sulfonamides were added.

Bootsma isolated the pathogen of CE in 1975. The characteristic features were described in 1977 [2]. Fijan made several experiments with the bacteria [4]. Bootsma et al. classified the pathogen as belonging to the *Aeromonas*, other researchers as belonging to the *Arthrobacter* genus [2, 7].

In the literature of fish diseases several studies have been published concerning the chronic dropsy syndrome [10]. A summary of the ulcerous form of the disease was given by Wunder in 1953 [12]. However, such research described infections caused by mixed bacteria only. As late as 1976 Amlacher still could not clearly determine the exact cause of the disease [1].

The chronic form of carp dropsy was described in our country by several authors [11]. Szakolczai made a summary report about bacteria that were the possible cause of the ulcerous form of the disease [11]. In 1976 Csaba and Békési isolated the *Aeromonas salmonicida* sp. obtained from ulcers of the carps and used it to induce infections artificially in the fish [3].

Materials and Methods

The observations were made on 50 one-year-old carp. The fish were kept in containers in 15°–20°C-water under favorable oxygen charge.

Broth cultures of the causative were applied to the scarified surface of the skin of experimental fish. At different intervals between 1 and 30 days histological sections were made of the pathological changes in dermal tissues. For this purpose samples were routinely collected in 10% formalin solution, for further procedure embedded in paraffin, hematoxilin-eosin, Ziehl-Neelsen, Giemsa-stainings loring, PAS reaction. Similarly samples were examined from a few internal organs in addition to the skin specimens.

---

1 Central Veterinary Institute, 1581 Budapest, Pf. 2, Hungary

Sensitivity tests with antibiotics and sulfonamide of various isolates were also carried out. Healing experiments were made on skin ulcers with enriched sulfonamide found to be beneficial.

Results

The 50 carp used in the experiments came from different reservoirs. Changes of varying degree appeared where the bacteria were rubbed into the ents made on the skin. The infection was most extensive on those spots where the cuts penetrated to deeper layers of the dermis and the watertemperature was at 20 °C. Other fish, with skin cuts only, were placed with the ones contaminated by bacteria. These showed the same characteristic skin lesions and development of the disease, though with some 6 days delay.

The pathogen could be found by bacterological examination in the ulcers only, but in more severe cases in the internal organs (spleen, kidney) too.

Histopathology

On the first day – by gross examination – the skin showed a slight redness only, which, microscopically, was found to be the result of the expanded veins of the dermis. On the second day the dermis became loose, the *Tonofibrillium* between the cells started to break up, and in addition to red cells a few *Polymorphonuclear leukocytes* showed up between the cells. The degeneration of the mucous cells of the cuts was followed by the degeneration of nondifferentiated cells of the epidermis. The veins in the dermis widened and focal hemorrhages could be observed. There appeared to be more of the polymorphonuclear leukocytes and less of the mononuclear type. This symptom did not extend into the stratum compactum, where only minor hemorrhages occurred. In the subdermis and in the muscular layer only expanded veins were observed (Figs. 1, 2).

From the 3rd to the 5th days characteristic knots showed up, manifesting the strong degeneration and thinning cut of the epidermis visible on histological sections. Contractions of the *Melanophores* were observed in the stratum terminans of the dermis. The edema of the dermis pushes the epidermis up in stages, creating a knot. Here the cell infiltration is slighter and the epidermis becomes extremely thin. At other spots around the veins small hemorrhages and mixed inflammatory cellular infiltrations are seen. This also extends into the stratum compactum and to the subdermis, and by aggravation of the process to the subdermal muscular layer beneath (Figs. 3, 4, 5, 6).

The period between the 11th and 15th days shows bloody-edged ulcers of irregular sizes, observed by the naked eye. Microscopically the ulcer is hemorrhagic.

Signs of regeneration in the epidermis appear as early as the 5th day, the epidermis acting as a limiting layer of tissue and consisting of inflammatory cells. The necrotic epidermis is replaced by a layer of fibroblasts which seals off the deeper inflamed area. At other places the process expands in depth and the polymorphonuclear cells are replaced by mononuclear ones in the epidermis. Melanophores located around take part in clearing up the residue of the tissue. The degenerated muscle fibers are replaced by

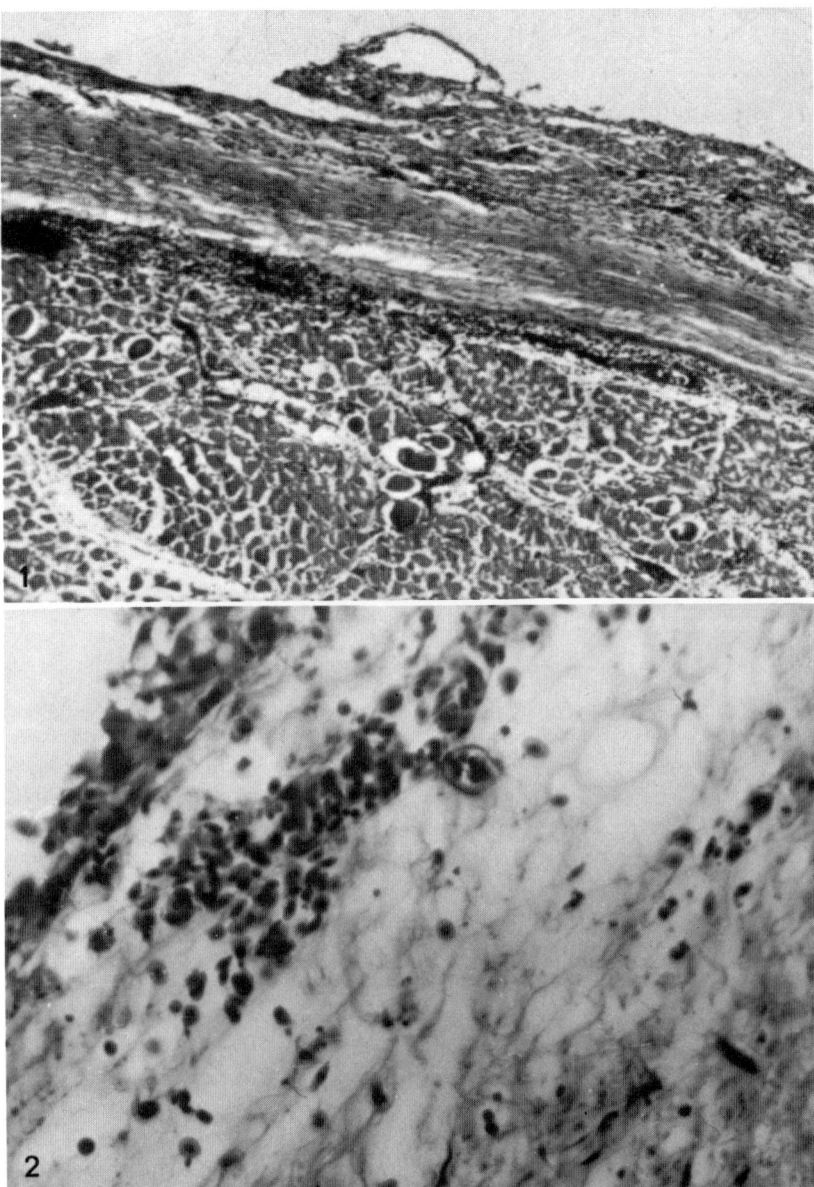

Fig. 1. CE infected carp. On the second day port. infection showing acute lesion: epidermal necrosis, edema, and hyperplasia of dermis. x 40. All figures: section stained with HE

Fig. 2. Inflammation, edema of dermis. x 400

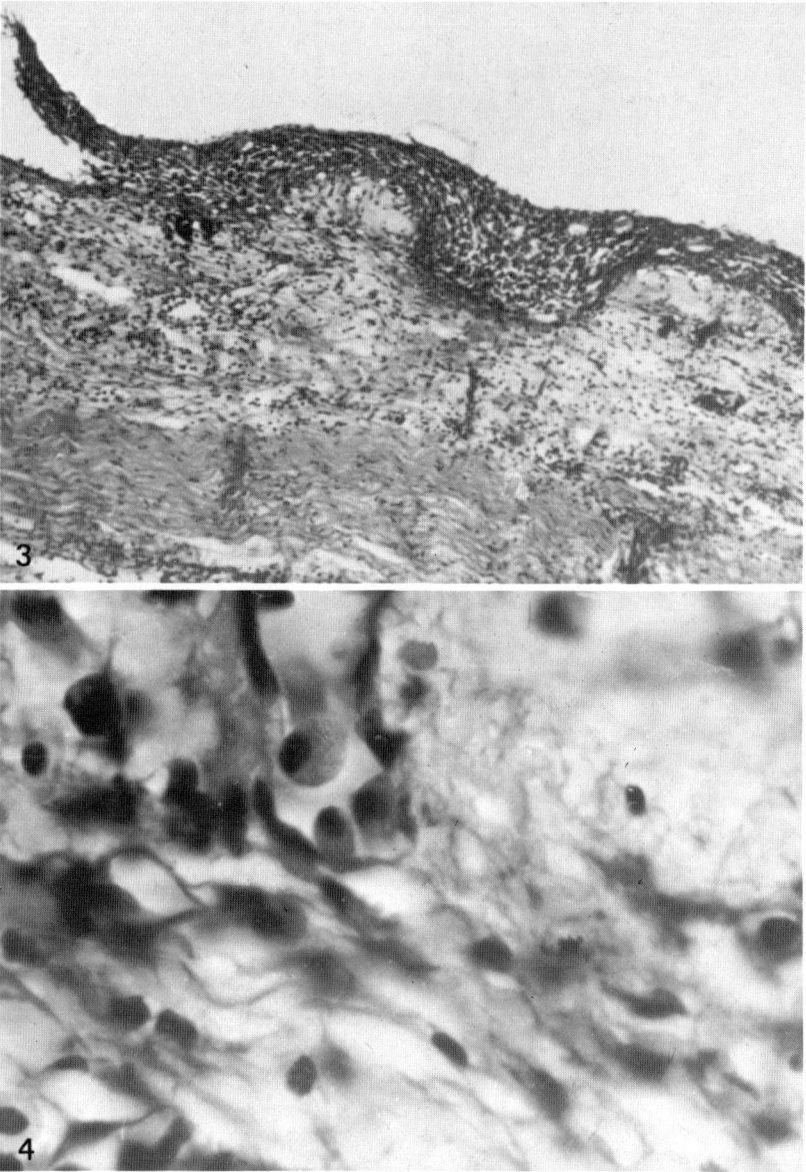

Fig. 3. On the 3rd to the 5th days, showing characteristic knots. x 100

Fig. 4. Edema of the knots of Fig. 3. x 400

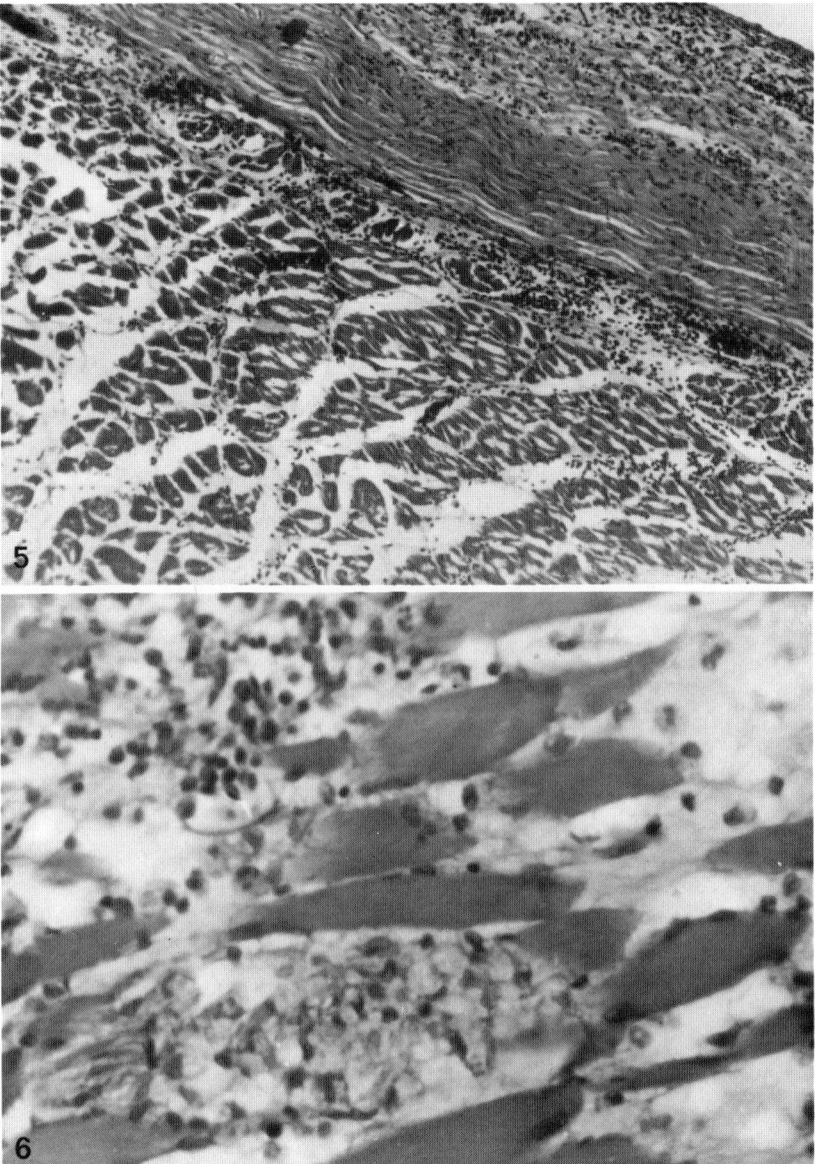

Fig. 5. Acute lesion showing inflammatory cellular infiltrations to the subdermis and to the subdermal layer. x 100

Fig. 6. Subdermal muscular layer inflammation, hemorrhage, and necrosis. x 400

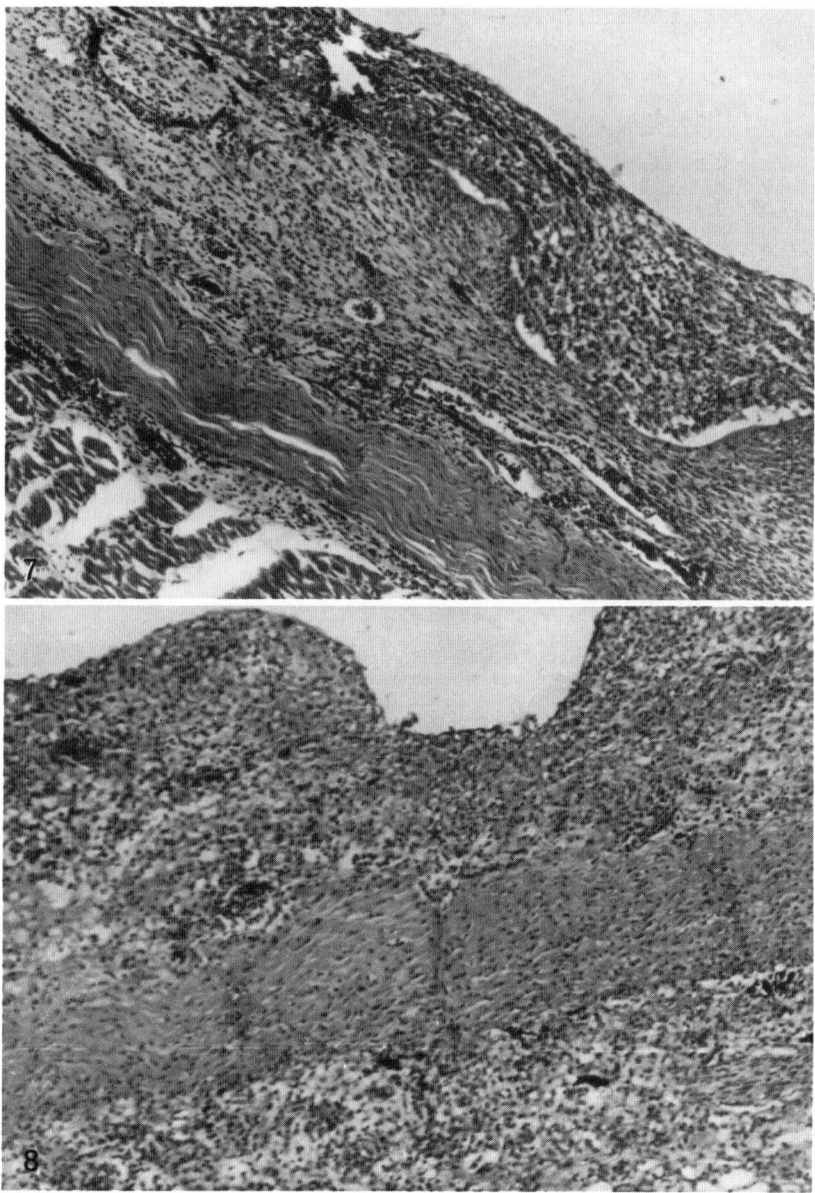

Fig. 7. Hemorrhagic ulcer. x 40

Fig. 8. More intense regeneration. x 40

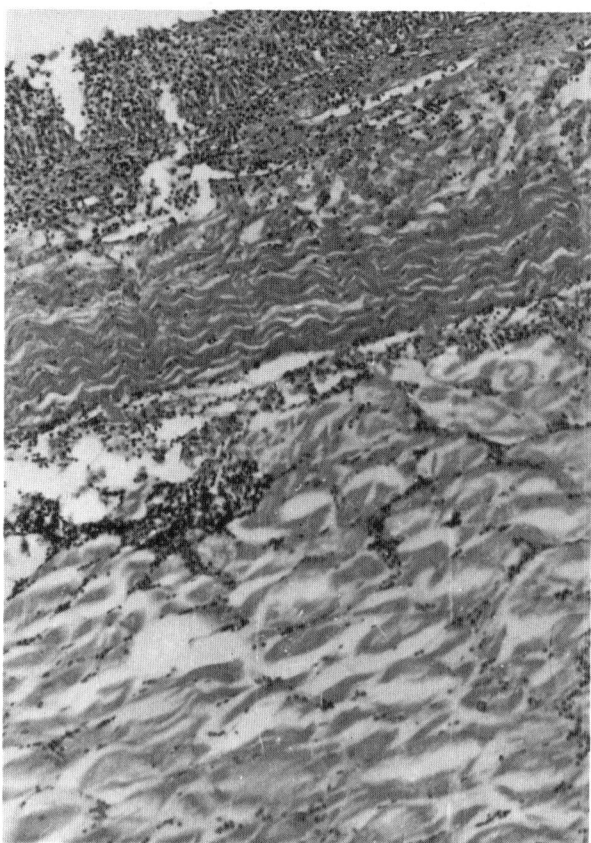

Fig. 9. Chronic inflammation: mononuclear cells infiltrate. x 40

growing fibers of connective tissue. From the 17th to the 30th days the healing ulcers become covered by epidermis (Figs. 7, 8).

Microscopic observation however reveals either further deterioration or a more intense regeneration of the deeper tissues. During this process the necrotic epidermis is cast off. The new thin epidermis does not contain any differentiated mucous cells. The change of the membrana basialis is not regular either. The melanophores are missing from the stratum laxum. The change of the fibers of the dermis in the stratum spongiosum does not comply with the control data. A great number of mononuclear cells infiltrate the deeper layers. The degenerated muscle fibers are replaced by fibers of the dermis (Figs. 9 and 10).

If there is no regeneration the process is followed by the recurrence of hemorrhagic inflammation, with a characteristic hemorrhagic septicemia at the end.

After the 5th day of the cuts the internal organs were examined. The macrophage centrums in the interlobular tissue of the liver and in the hemopoetic tissue of the kidney and the spleen showed increased activity. Certain more serious changes which were experienced in small numbers only are not mentioned here due to possible sources of errors.

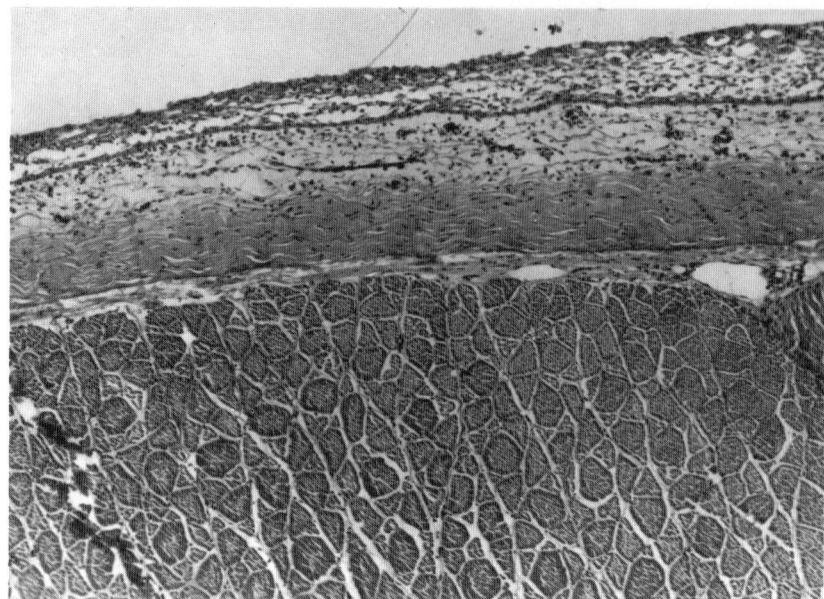

Fig. 10. The healing of the skin lesion. x 40

Treatment of the developed ulceral changes was done by the combination of sulfonamide and trimethoprim. After two applications of the medicine acute inflammation became chronic. The disease did not go deeper. The magnitude of the infiltration of the mononuclear cells increased. No secondary infection became evident. The ulcers completely regenerated on the 6th day. In more advanced cases the healing percentage is not as satisfactory.

Discussion

By the pathological examinations the processes of skin inflammation and inflammatory necrosis caused by the pathogen of CE could be followed.

These experiments showed that the pathogen could not penetrate the intact epidermis. For this reason skinscarification was used, as such lesions could frequently occur in our reservoirs under natural circumstances. Various parasites, the Lernaea species in the first place, may have the same role as artificial scarification in inducing skin lesions.

The artificial infection was most effective when the pathogen penetrated the dermis. The change was of the highest percentage when the water temperature was kept at 20 °C. This could be the reason that the natural acquisition of the disease in our climate can be observed for 5–9 months.

The process caused by the CE pathogen develops as follows. It starts with extensive hemorrhages in the dermis. The inflammation begins with infiltration of a few inflammatory cells. This is followed by the degeneration and necrosis of the epidermis.

The edematic serous infiltration lifts up the epidermis by knots, meanwhile the epidermis becomes thin, dies, and an ulcer develops. The inflammatory cellular infiltration expands from the stratum spongiosum to the stratum compactum, and then to the subdermis and the muscles. Up to this point we can speak of an initial acute inflammation of the skin. The healing process starts on the 5th day after the infection but becomes more pronounced on the 11th day. The proportion of inflammatory cells shifts more and more to the mononuclear type. After the ulcer is sealed we can speak of a chronic inflammation of the skin. The disease observed can become more severe depending upon the general resistance of the body under unfavorable external circumstances, or it can heal slowly under advantageous aquatic conditions. The healing process can be supported by medicaments. The combination sulfonamide-trimethoprim puts the healing process on the right course.

The experimentally induced changes of the skin were compared with ones which were acquired naturally, with particular reference to the causative agents. It was established that pathological changes acquired under natural conditions were similar to those produced artificially. Especially in the initial stage of changes consequences can be drawn concerning the results of our examinations of affected tissue and the time of development of the disease. The picture may be somewhat distorted by the presence of secondary pathogens or parasites. When the pathological changes of the skin are caused by Lernaea sp. they are characterized by hyperplasia of the epidermis, edema of the dermis associated with a great number of eosinophils. Later on in the process an encasement around the necrotic spot can be observed. All these characteristics set our examinations apart from disease caused by *Aeromonas salmonicida* spp. The *Argulus* sp. often found in the skin of the carp results in a characteristic hypertrophy of the epidermis and may create hemorrhages, too. More investigations are needed to define additional possible roles of other pathogens in causing the worsening of the skin ulcer in carp.

Histopathological lesions due to the causative agent of CE have not been described so far. Wunder [12] in his paper on chronic dropsy described histological details of alterations known as mixed infection at that time. Dermal changes such as formation of the initial stage of papilla growth to the development of ulcers can be followed in CE. In cases caused by *Aeromonas salmonicida* species, according to our experience, it is characteristic that a definite papilla growth in the dermis and ulceral formations are seen as consequences of microscopic abscesses. Huzinga et al. [5] were successful in producing skin lesions artificially in other species of fish with *Aeromonas hydrophyla* strains. These dermal changes were rather similar to our histological picture due to CE agent, but in the latter there was no severe necrosis in the internal organs.

According to Mackie and Menzies [8] typical vesicular lesions and ulcers in the skin can be produced by parenteral application of the causative agent of furunculosis in fish. As far as this pathogen is concerned it is most significant that – as Klontz et al. stated [6] – leukocytes appeared in the skin lesions after 48 h. The same was the case in our experiments, too. There follows a leukopenia that we also saw in CE. In Amlacher's opinion the pathological process of furunculosis usually begins in the subdermis, while in CE it always starts from the dermis. But it is obvious that there are a lot of similarities in the development of papilla and ulcers of both diseases.

Finally it can be stated that artificial infections by a well-known pathogen are of high value in differential diagnosis, and in the control of diseases too.

References

1. Amlacher E (1976) Taschenbuch der Fischkrankheiten. G. Fischer, Jena
2. Bootsma R, Fijan N, Blommaert J (1977) Isolation and preliminary identification of the causative agent of carp erythrodermatitis. Vet Arch 47:291–302
3. Csaba GY, Békési L, Körmendy B (1978) Pontyok bőrelváltozásából izolát baktériumok vizsgálata. MAL 33:X72
4. Fijan N (1973) Carp erythrodermatitis – a review. EIFAC/T 17:Suppl 2:113–118
5. Huzinga HW, Esch GW, Hazen TC (1979) Histopathology of red-sore disease/Aeromonas Hydrophila in naturally and experimentally infected largemouth bass Micropterus salmoides, Lecépède. J Fish Dis 2:263–277
6. Klontz GW, Yasutake WT, Johnross A (1966) Bacterial disease of the Salmonidae in the western United States: Pathogenesis of furunculosis in rainbow trout. Am J Vet Res 27:455
7. Kölbl O (1978) Bakterien als Ursache der Erythrodermatitis der Karpfen/Chronische Form der IBW. Österr Fisch 31.11/12:201–205
8. Mackie TH, Menzies WJM (1938) Investigations in Great Britain of furunculosis of the Salmonidae. J Comp Pathol Ther 51:225
9. Roberts RJ (1978) Fish pathology. Bailliere Tindall, London
10. Schäperclaus W (1965) Etiology of infectious carp dropsy. Ann Acad Sci 126:587
11. Szakolczai J (1969) Adatok a heveny hasvizkór oktanához és kórfejlödéséhez – Kandidátusi disszertáció. Budapest
12. Wunder W (1953) Makroskopische und mikroskopische Pathologie der infektiösen Bauchwassersucht des Karpfens. Z Fisch 5/6:335–357

Erythrodermatitis of Carp: Studies of the Mode of Infection

D. SCHULZ[1]

Erythrodermatitis is characterized by inflammatory lesions of the skin and by necrosis leading to ulceration. The disease is common in the majority of carp fish farms in Europe (Fijan, 1972). It is a contagious disease of the skin of carp with varying severity and lethality. The etiology of this disease has not yet been established. In spite of the fact that a number of scientists has studied this disease, there are still problems (Schäperclaus, 1930, 1967; Schäperclaus and Mann, 1939; Schubert, 1960, 1962, 1963, 1964 a,b, 1967; Brunner, 1961; Heuschmann-Brunner, 1965, 1970, 1971, 1978; Bootsma et al., 1977). An outbreak of clinical erythrodermatitis in a carp pond in West Germany was reported to our laboratory within the Division of Zoonoses and Epizootics in 1976. A group of 67 two-year-old mirror carps from this farm pond were examined bacteriologically. The pond carps were killed electrically, their skin branded with a red-hot iron, and incisions made with sterile scissors in order to obtain material from the following organs: heart, liver, kidney, swim bladder, gill. Material from the ulcers was not taken because it was thought to be contaminated by other water bacteria. A loopful of the tissue fluid of sulfide) medium incubated at 30 °C for 24 h, nutrient agar and trypticase soy agar incubated at 22 °C for 48 h]. Colonies in pure culture were isolated and identified by the bacteriological methods described in Bergey's Manual (1974). Data on the isolates are given in Table 1. In total, 92 strains were isolated. From these, 45 *Aeromonas* strains belonged to the *hydrophila punctata* group, 27 strains to the genus *Pseudomonas*, and 20 strains were from other bacterial species.

For pathogenicity tests, mirror carps weighing approximately 60 g each were used. These fish had been kept in aquaria or in laboratory-owned ponds for several months without showing any clinical signs of disease. The aquaria were continuously supplied with tap water of 20 °C. The carps were automatically fed with commercial pelleted trout food 5 times during 24 h. Before experimental infection, the carps were narcotized electrically and scarified with a flamed vaccinostyle that had been dipped into the culture medium. The unscaled skin was scarified by perforating the epidermis but not the dermis.

In the first pathogenicity test involving different strains of bacteria that had been isolated, 24-h culture medium from blood agar was used. After 3 months had passed and no clinical signs of erythrodermatitis were observed, experimental conditions were changed: the water temperature was lowered to 12 °C and the fish put under stress

---

[1] Division of Zoonoses and Epizootics, Institute for Veterinary Medicine, Federal Health Office, Postfach, D-1000 Berlin 33

Table 1. Isolation of bacteria from a clinical outbreak of erythrodermatitis

Isolated bacteria from the organs of diseased fish	92 strains
Aeromonas:	*45 strains*
(*hydrophila punctata* group)	
A. hydrophila subsp. *hydrophila* biotype I	10 strains
A. hydrophila subsp. *hydrophila* biotype II	19 strains
A. hydrophila subsp. *anaerogenes* biotype I	3 strains
A. hydrophila subsp. *anaerogenes* biotype II	3 strains
A. punctata subsp. *punctata*	4 strains
A. punctata subsp. *caviae*	4 strains
A. VPR-neg. *anaerogenes*	3 strains
Pseudomonas:	*27 strains*
P. maltophila	1 strain
P. stutzeri	1 strain
P. alcaligenes	2 strains
P. diminuta	4 strains
P. vesicularis	3 strains
P. cepacia	1 strain
P. putrefaciens	15 strains
Vibrio:	
V. alginolyticus	3 strains
Plesiomonas:	
P. shigelloides	1 strain
Brucella:	
Pasteurella	3 strains
Moraxella	2 strains
Flavobacterium	2 strains
Achromobacter Mucosus Group:	2 strains
Enterobacteriacea:	
Citrobacter:	1 strain
Klebsiella:	1 strain
Enterobacter:	1 strain
S. liquefaciens:	3 strains
Proteus providencia group:	1 strain

by administration of prednisolone acetate in a dosis of 20 mg/kg body weight. The fishes were again infected with bacteria by scarification. Two months after infection, the fish did not show any effect except a lower body weight than the controls. Therefore, the method of infection was changed again: a flamed dissection needle was thrust into the skin (Fig. 1) between epidermis and dermis to prepare a channel in the tissue. By means of a straight platinum wire pure culture material was introduced into this channel. In addition, the opening of the channel was closed with Histoacryl tissue plaster, which is used for kidney and liver operations because it is resistant to fluids.

Fig. 1. Mirror carp experimentally infected with *Klebsiella* (strain 884 I), showing erythrodermatitis lesions (a hemorrhagic inflammatory process develops between the epidermis and the dermis); picture taken 4 days after infection. Infection was performed by a flamed dissection needle which was thrust into the skin to prepare a channel between epidermis and dermis. With a straight platinum wire, pure culture material was introduced into the channel. After this the channel was closed with Histoacryl tissue plaster

The culture medium was also changed and a mixture of tryptose (1%), serum (10 vol.%) and agar (1.2%) was used. Mortality was recorded every 24 h. Reisolation of all strains was performed approximately 10 days after infection using the method and isolation technique described before.

In spite of the fact that now the fish showed symptoms of erythrodermatitis in several cases, often other bacterial species were found. It was even not quite established that when *Aeromonas hydrophila* was isolated, these bacteria were identical with those that had been inoculated into the fish. Therefore, the bacteria were rendered resistant with nalidixic acid before infecting the fish. Not all of the bacteria used could be made resistant: 19 strains did not become resistant and thus could not be used in the experiments. The results of pathogenicity testing are presented in Table 2. Mortality was highest in carps infected with *Aeromonas hydrophila* subspecies *hydrophila* biotope II. The ulcers which occurred were not always found on the site of infection. Sometimes they were situated on the opposite side of the body and sometimes on the abdomen. Very severe symptoms were caused by *Aeromonas* subspecies *anaerogenes* biotope I and by anaerogenic aeromonads in the Voges-Proskauer test. In the cases first an extended inflammation was observed, which was followed by an extended necrosis. From the genus *Pseudomonas*, *P. putrefaciens* was identified as the pathogen in 24% of the cases, causing symptoms of erythrodermatitis and mortality. *Vibrio alginolyticus* was also pathogenic for mirror carps. *Klebsiella* was highly pathogenic for carps and *Serratia liquefaciens* gave only a weak reaction. No pathological changes were caused by *Pseudomonas maltophila*, *P. alcaligenes*, *P. diminuta*, *Plesiomonas shigelloides*, *Moraxella*, *Achromobacter mucosus* group, *Enterobacter*, and *Proteus providencia* group (see Figs. 2–6).

Table 2. Results of infection experiments

Strain	No. of infected fishes	Mortalities	Symptom of erythrodermatitis
A. hydrophila subsp. hydrophila biotype I	20	6	13
A. hydrophila subsp. hydrophila biotype II	59	22	37
A. hydrophila subsp. anaerogenes biotype I	8	7	5
A. hydrophila subsp. anaerogenes biotype II	9	5	6
A. punctata subsp. punctata	8	1	6
A. punctata subsp. caviae	7	1	1
A. VPR-neg. anaerogenes	7	2	5
Pseudomonas maltophila	6	–	–
P. stutzeri	6	–	2
P. alcaligenes	6	–	–
P. diminuta	6	–	–
P. vesicularis	6	–	3
P. cepacia	6	–	3
P. putrefaciens	37	7	8
Vibrio alginolyticus	6	–	3
Plesiomonas shigelloides	7	–	–
Pasteurella	11	–	1
Moraxella	6	–	–
Flavobacterium	7	1	2
Achromobacter mucosus group	6	–	–
Citrobacter	6	–	3
Klebsiella	6	–	4
Enterobacter	6	–	–
Serratia liquefaciens	6	–	5
Proteus providencia group	6	–	3
Ref. strain: Bootsma V 76/134	6	2	6
Ref. strain: Soc. Microbiol. Göttingen	6	–	6
Ref. strain: Animal Health Serv. Grub	6	–	6

Comparative studies with reference strains were made using an *Aeromonas hydrophila* strain from the Society of Microbiology in Göttingen, an "erythrodermatitis" strain from Fijan/Bootsma (V 76/134) and two "erythrodermatitis" strains from the Fish Health Service, Grub (445/78 and 314/78). Under the same experimental conditions the reference strains showed a low pathogenicity.

In conformity with the findings of Heuschmann-Brunner (1978), most of the isolated strains belonged to the *Aeromonas hydrophila* group. This group is found not only in diseased fish but also in fish without symptoms of a disease, in surface water,

Fig. 2. Mirror carp experimentally infected with *Klebsiella* (strain 884 I), showing erythrodermatitis lesions; picture taken 4 days after infection. Experimental infection was performed by methods of skin scarification. The skin was scarified by perforating the epidermis but not the dermis

Fig. 3. Mirror carp experimentally infected with *Aeromonas* VPR-negative anaerogenic (strain 878 I) showing erythrodermatitis lesions (the inflammatory zone gradually extends; tissue breakdown leads to the formation of an ulcer with central necrosis); picture taken 3 days after infection. Infection was performed by the method mentioned in Fig. 1

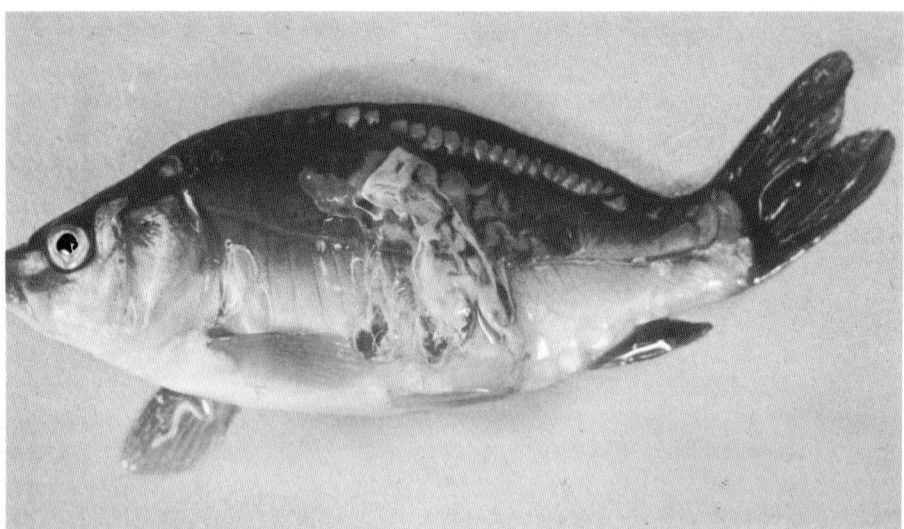

Fig. 4. Mirror carp experimentally infected with *Aeromonas* VPR-negative anaerogenic (strain 878 I) showing erythrodermatitis lesions as in Fig. 3. Infection was performed by the method mentioned in Fig. 2. Picture taken 3 days after infection

Fig. 5. Mirror carp experimentally infected with *Proteus providencia* group (strain 875 I) showing erythrodermatitis lesions (a hemorrhagic inflammatory process develops between the epidermis and the dermis); picture taken 4 days after infection. Infection was performed by the method mentioned in Fig. 1

Fig. 6. Mirror carp experimentally infected with *Proteus providencia* group (strain 875 I) showing erythrodermatitis lesions; picture taken 4 days after infection. Infection was performed by the method mentioned in Fig. 2

and in sewage. In polluted water, the number of *Aeromonas* may amount to between several thousand and 10,000 bacteria per milliliter of water. Therefore, it is troublesome to estimate the infectivity which is transmitted from the water to the fish. The fish-pathogenic strains of *A. hydrophila* subspecies *hydrophila* in Schubert's collection had been isolated from surface water.

Our experiments have shown that different species of bacteria may cause the symptoms of erythrodermatitis and that for their pathogenicity the culture media and the route of experimental infection are important. The method of scarification is inexact because a number of bacteria may be washed out into the surrounding water or, vice versa, may settle on the injured skin and thus mask the true character of the infection. On the other hand, there are different possibilities for the fish to get injuries of the epidermis, by ectoparasites, or by detergents, etc. These injuries may lead to openings in the skin, providing portals of entry for the bacteria. A number of diseases of freshwater fish is due to complexes of responsible injurious factors. A use of SPF fish in experiments to elucidate the etiology of this disease would be most helpful.

Acknowledgements: I would like to thank Prof. Dr. Bulling, Chief of the Division of Zoonoses and Epizootics, Federal Health Office Berlin for supporting these studies and Mrs. Gabriele Schulz and Mrs. Eleonore Rademacher for their very skilful technical assistance. I am also grateful to Dr. Bootsma, Utrecht and Dr. Wiedemann, Grub for providing reference strains.

References

Bergey's (1974) Manual of determinative bacteriology, 8 edn. Williams & Wilkins Comp, Baltimore

Bootsma R, Fijan N, Blommaert J (1977) Isolation and preliminary identification of the causative agent of carp erythrodermatitis. Vet Arch 47:291–302

Brunner G (1961) Neuere Erkenntnisse über die infektiöse Bauchwassersucht des Karpfens. Fischwirt 11:240–241

Fijan NN (1972) Infectious dropsy of carp – a disease complex. In: Mawdesley-Thomas LE (ed) Diseases of fish. Academic Press, London New York, pp 39–51

Heuschmann-Brunner G (1965) Ein Beitrag zur Erregerfrage der infektiösen Bauchwassersucht des Karpfens. In: Festschrift: Der Fisch in Wissenschaft und Praxis. Sonder Allg Fischz 90: 41–49

Heuschmann-Brunner G (1970) Die Aeromonaden in der Hydrobiologie. Z Wasser Abwasser Forsch 3:40–41

Heuschmann-Brunner G (1971) Einige Bemerkungen aus bakteriologischer Sicht zu dem Thema „Infektiöse Bauchwassersucht des Karpfens". Münch Beitr Abwasser Fisch Flussbiol 20:67–69

Heuschmann-Brunner G (1978) Die Aeromonaden der „Hydrophila-Punctata-Gruppe" bei Süßwasserfischen. Arch Hydrobiol 83:99–125

Schäperclaus W (1930) Pseudomonas punctata als Krankheitserreger bei Fischen. Untersuchungen über Süßwasseraalrotseuche, Leibeshöhlenwassersucht der Cypriniden, insbesondere des Karpfens und Fleckenseuche der Weißfische. Z Fisch 28:289–370

Schäperclaus W (1967) Probleme der Karpfenimmunität gegenüber Aeromonas punctata und Fragen der antigenen Struktur des Bakteriums. Z Fisch NF 15:129–138

Schäperclaus W, Mann H (1939) Untersuchungen über die ansteckende Bauchwassersucht des Karpfens und ihre Bekämpfung. Z Fisch 37:1–182

Schubert RHW (1960) Untersuchungen über die Merkmale der Gattung Aeromonas. Zentralbl Bakteriol I Orig 180:310–327

Schubert RHW (1962) Über die biochemischen Eigenschaften der anaerogenen Aeromonaden. Zentralbl Bakteriol I Orig 185:503–511

Schubert RHW (1963) Über die biochemischen Eigenschaften von Aeromonas hydrophila. Zentralbl Bakteriol I Orig 188:62–69

Schubert RHW (1964a) Zur Taxonomie der Voges-Proskauer-negativen „hydrophila-ähnlichen" Aeromonaden. Zentralbl Bakteriol I Orig 193:482–490

Schubert RHW (1964b) Zur Taxonomie der anaerogenen Aeromonaden. Zentralbl Bakteriol I Orig 193:343–352

Schubert RHW (1967) Das Vorkommen der Aeromonaden in oberirdischen Gewässern. Arch Hyg 150:688–708

Emerging Problems and Approaches

Chairman: P. GHITTINO, Italy

Characterization of the Causal Agents of Bacterial Kidney Disease

B. AUSTIN[1]

Introduction

Bacterial kidney disease (BKD) is a severe debilitating disease of salmonid fish, particularly *Oncorhynchus* spp. and *Salmo* spp., with incidences in North America (Ordal and Earp, 1956; Bullock et al., 1974), Japan (Kimura, 1978), and Europe, particularly France (Evelyn, 1977), Great Britain (Smith, 1964), Turkey (Halici et al., 1977), and Yugoslavia (Fijan, 1977). The aetiological agent has been misclassified, although it has been referred to as a coryneform or *Corynebacterium* (Bullock et al., 1974; Vladik et al., 1974; Sanders et al., 1978), on account of the distinctive micro-morphology of small gram-positive, paired cocco-bacilli. However, the genus *Corynebacterium* contains many asporogenous gram-positive rods, and extensive efforts have been made to re-define the genus. Thus, *Corynebacterium* should be restricted *sensu stricto* to near neighbours of the human pathogen, *C. diphtheriae*, i.e., club-shaped cells with evidence of snapping division (Cowan, 1974; Minnikin et al., 1978). Hence, the majority of BKD isolates do not conform to the true definition of *Corynebacterium*.

Diagnosis of BKD has resulted usually from histological examination of kidney, and the isolation of slow-growing gram-positive, rod-shaped bacteria on enriched media (Ordal and Earp, 1956), no further effort being made to identify the pathogen. Consequently, it has been the purpose of this study to characterize these micro-organisms in depth.

Materials and Methods

Source of Strains

Twenty-one strains, each considered to be a causal agent of BKD were received from several sources (Fig. 1). Four fresh UK isolates, 1/79, 2/79, 10/79, and 11/79, were obtained from swabs of kidney removed aseptically from diseased rainbow trout *(Salmo gairdneri)* and inoculated on to plates of Mueller-Hinton agar (Oxoid) supplemented with 0.1% (w/v) L-cysteine hydrochloride and 20% (v/v) fetal bovine serum. Plates were examined after 28 days incubation at 16 °C, when isolates were removed, and re-streaked three times on fresh media to ensure purity. All strains were maintained on modified Mueller-Hinton agar at 4 °C.

[1] Ministry of Agriculture, Fisheries and Food, Directorate of Fisheries Research, Fish Diseases Laboratory, Weymouth, Dorset DT 4 8UB, England

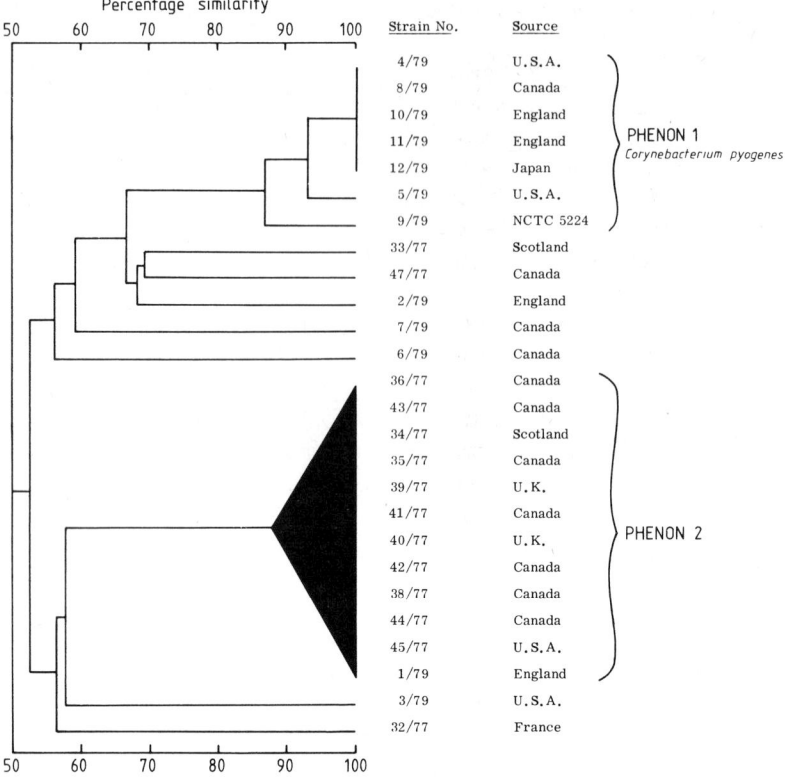

Fig. 1. A simplified dendogram based on the S_J coefficient and single-linkage method of clustering, showing relationships between the strains

Characterization of the Strains

Each strain was examined for over 100 unit characters, as currently used in extensive taxonomic investigations (Colwell and Weibe, 1970), using modified Mueller-Hinton agar or media supplemented with 0.1% (w/v) L-cysteine hydrochloride and 5% (v/v) fetal bovine serum as the basal medium. Tests were recorded after prolonged incubation at 16 °C for 28 days, when results were reduced to a unit character state, and the data examined by the procedures of numerical taxonomy (Sneath and Sokal, 1973).

Electron Microscopy

Cells grown on basal medium were suspended in 0.9% (w/v) saline and negatively stained with 2% (w/v) aqueous uranyl acetate. The micro-morphology was determined using a Jeol CX100 transmission electron microscope.

Guanine Plus Cytosine Content of DNA

Despite numerous attempts, purified DNA could not be isolated from cells of 34/77, 39/77, 5/79 or 8/79.

Serology

Washed cells were suspended in sterile 0.9% (w/v) saline to ca. 10^8 cells/ml and their agglutination reactions were determined using doubling dilutions of antisera to strains 33/77, 36/77, 39/77, 41/77, and 2/79, prepared in female New Zealand white rabbits (McCarthy and Rawle, 1975).

Results and Discussion

On the basis of overall similarity, the strains separated into 2 numerically dominant groups, phena 1 and 2, and 7 single member clusters (Fig. 1). Phenon 1 contained the type strain of *Corynebacterium pyogenes* (NCTC 5224).

Characteristics of the Strains

A summary of the characteristics of phena 1 and 2 has been given in Table 1. With the exception of strains 6/79 and 7/79, which comprised large gram-positive rods, all isolates appeared morphologically similar when gram-stained smears and wet preparations were examined by light microscopy. However, subtle variation in cell size and shape were revealed by high-resolution transmission electron microscopy. Thus strain 5/79, a representative of phenon 1, consisted of diplo-cocco-bacilli, 2.0x1,8 μm in size (Fig. 2), whereas strain 34/77 of phenon 2 comprised rods, 3.0x1.0 μm, occurring singly (Fig. 3) or, rarely, in pairs. Moreover, distinctions between phena extended beyond micro-morphology insofar as the 6 strains of phenon 1 were extremely fastidious in their growth requirements, taking 28 days at 16 °C to produce colonies on enriched Mueller-Hinton agar. In contrast, members of phenon 2 grew readily on a wide assortment of standard bacteriological media, including nutrient agar (Oxoid) and tryptone soya agar (Oxoid). Thus, from a comparison of the phenotypic characteristics, it would appear that the micro-organisms considered as the causal agents of BKD, represent a diverse spectrum of bacterial taxa. Nevertheless, serology data summarised in Table 2 show that there was cross-agglutination between antisera and strains representative of phena 1 and 2, and some of the unclustered isolates. Therefore, it is apparent that there are some shared antigens among the isolates.

Taxonomic Considerations

The presence of the type culture of *Corynebacterium pyogenes* (NCTC 5224) with isolates clustered in phenon 1 has, on first appearances, provided much information

Table 1. Summary of the characteristics of the isolates clustered in phena 1 and 2

Characteristic	Phenon 1	Phenon 2
Biochemistry:		
Aesculin degradation	−	+
Arginine dihydrolase	−	+
Casein hydrolysis	−	+
Catalase	(+)[a]	(+)[a]
Methyl-red test	−	+
Phosphatase	−	+
Tween 40 hydrolysis	−	+
Growth in:		
2.5% (w/v) NaCl	−	+
Enriched CO_2	+	+
Anaerobiosis	+	+
Crystal violet	−	+
Malachite green	−	+
Potassium tellurite	+	+
Growth at:		
4 °C	−	+
16 °C	+	+
22 °C	−	+
Sensitivity to:		
Ampicillin	+	+
Cephaloridine	+	+
Cloramphenicol	+	+
Erythromycin	+	−
Tetracycline	+	+

[a] Weakly positive

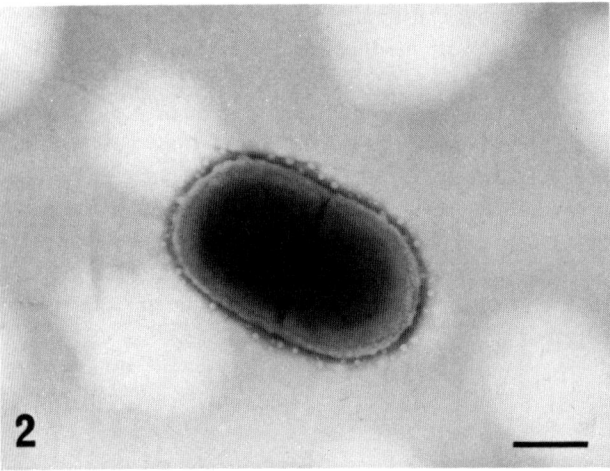

Fig. 2. Negatively stained cell representative of the isolates clustered in phenon 1. *Bar* equals 1 μm

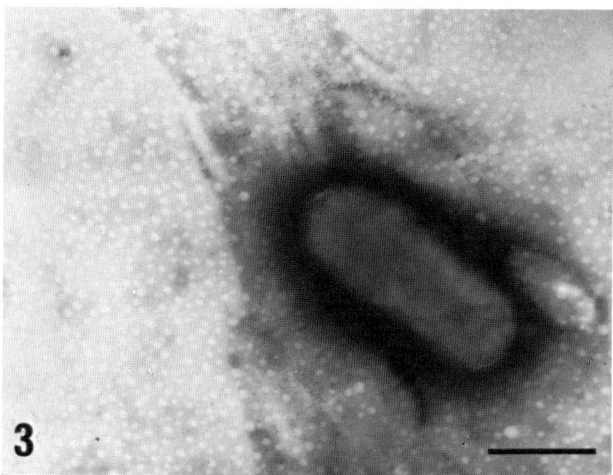

Fig. 3. Transmission electron micrograph of cell representative of isolate from phenon 2. *Bar* equals 1 μm

on the taxonomic status of this group of micro-organisms. Moreover, there was very close similarity between the description of *C. pyogenes* (Cummins et al., 1974) and the characteristics of phenon 1, except for the inability of the fish isolates to grow at 37 °C. Unfortunately, *C. pyogenes* has been considered distinct from other *Corynebacterium* spp. (Jones, 1975) to such an extent that it has been proposed to accommodate the species in another, as yet unspecified, genus (Harrington, 1966; Minnikin et al., 1978). The relationship of this taxon to the lactic acid bacteria, notably *Streptococcus* and *Lactobacillus,* has been indicated (Whittenbury, 1964; Cummins et al., 1974), and needs clarification. However, it is noteworthy that the electron micrograph of 5/79 (Fig. 2) is indicative of the spherical cells of *Streptococcus*. Thus, the classification of phenon 1 must await improvements in the taxonomy of *C. pyogenes*.

Table 2. Titres obtained by agglutination of representative BKD strains against antisera

Bacterial isolate		Antisera to:				
		36/77	39/77	41/77	33/77	2/79
Phenon 1	4/79	1/64	1/64	1/128	1/1,024	1/2,048
	8/79	1/64	1/64	1/128	1/1,024	1/2,048
Phenon 2	34/77	1/32	1/512	1/128	–	1/8
	36/77	1/1,024	1/128	1/1,024	1/8	1/4
	1/79	1/128	1/2,048	1/1,024	1/4	1/4
Miscellaneous	32/77	1/8	1/8	1/4	1/8	1/8
	2/79	–	–	1/32	–	1/128
	3/79	1/2	1/16	1/2	1/2	–
	6/79	1/512	1/8	1/32	1/16	1/16

Similarly, the taxonomic status of phenon 2 is unclear. With *Corynebacterium* restricted *sensu stricto* to near neighbours of *C. diphtheriae,* clearly phenon 2 belongs in another genus. *Lactobacillus* may be an appropriate home for these organisms, insofar as they meet the genus description of gram-positive, non-motile, catalase-negative, fermentative rods with complex organic nutritional requirements (Rogosa, 1974). However, with the guanine plus cytosine content of DNA occupying a range of 34.7 to 53.4 mole% (Rogosa, 1974), this genus is in dire need of taxonomic revision. It is noteworthy that *Lactobacillus* was recognised previously as pathogenic to fish, causing "pseudokidney" disease (Ross and Toth, 1974), although the characteristics of this micro-organism differ considerably from those of phenon 2.

Therefore, the strains examined in this study have been shown to be related to the lactic acid bacteria, notably *Lactobacillus* and *Streptococcus*. However, it is disquieting that such a wide diversity of bacterial taxa should be associated with one disease. Perhaps the vague description of BKD is at fault. Thus inexperienced workers may be confusing several chronically related diseases, or even isolating erroneous strains on laboratory media. Nevertheless, it is apparent that more work is necessary to clear the myths enshrouding BKD.

The reference to proprietary products in this paper should not be construed as an official endorsement of these products, nor is any criticism implied of similar products which have not been mentioned.

References

Bullock GL, Stuckey HM, Chen PK (1974) Corynebacterial kidney disease of salmonids: growth and serological studies on the causative bacterium. Appl Microbiol 28:811–814

Colwell RR, Weibe WJ (1970) "Core" characteristics for use in classifying aerobic heterotrophic bacteria by numerical taxonomy. Bull Ga Acad Sci 28:165–185

Cowan ST (1974) Cowan and Steel's manual for the identification of medical bacteria. 2nd edn. Cambridge University Press, Cambridge, pp 238

Cummins CS, Lelliott RA, Rogosa M (1974) Genus 1. *Corynebacterium* Lehmann and Newmann 1896, 350. In: Buchanan RE, Gibbons NE (eds) Bergey's manual of determinative bacteriology, 8th edn. Williams and Wilkins Co, Baltimore, pp 602–617

Evelyn TPT (1977) An improved growth medium for the kidney disease bacterium and some notes on using the medium. Bull Off Int Epiz 87:511–513

Fijan N (1977) Corynebacteriosis (Dee disease, kidney disease) of salmonids in Yugoslavia. Bull Off Int Epiz 87:509

Halici G, Istanbulloglu E, Arda M (1977) An outbreak of bacterial kidney disease in fish farming station of Bayinder Dam and its treatment (in Turkish). J Fac Vet Med Univ Istanbul 3:22–27

Harrington BJ (1966) Numerical taxonomic study of some corynebacteria and related organisms. J Gen Microbiol 45:31–40

Jones D (1975) A numerical taxonomic study of coryneform and related bacteria. J Gen Microbiol 87:52–96

Kimura T (1978) Bacterial kidney disease of salmonids. Fish Pathol 16:43–52

McCarthy D, Rawle CT (1975) The rapid serological diagnosis of fish furunculosis caused by "smoth" and "rough" strains of *Aeromonas* salmonicida. J Gen Microbiol 86:185–187

Minnikin DE, Goodfellow M, Collins MD (1978) Lipid composition in the classification and identification of coryneform and related taxa. In: Bousfield IJ, Calleby AG (eds) Coryneform bacteria. Academic Press, London New York, pp 85–160

Ordal EJ, Earp BJ (1956) Cultivation and transmission of etiological agent of kidney disease in salmonid fishes. Proc Soc Exp Biol Med 92:85–88

Rosoga M (1974) Genus 1. *Lactobacillus* Beijerinck 1902, 212, Nom cons Opin 38, Jud Comm 1971, 104. In: Buchanan RE, Gibbons NE (eds) Bergey's manual of determinative bacteriology, 8th edn. Williams and Wilkins Co, Baltimore, pp 576–593

Ross AJ, Toth RJ (1974) Lactobacillus – a new fish pathogen. Prog Fish Cult 36:191

Sanders JE, Pilcher KS, Fryer JL (1978) Relation of water temperature to bacterial kidney disease in coho salmon *(Oncorhynchus kisutch)*, sockeye salmon *(O. nerka)*, and steelhead trout *(Salmo gairdneri)*. J Fish Res Board Can 35:8–11

Smith IW (1964) The occurrence and pathology of Dee disease. Freshwater Salmon Fish Res 34:1–12

Sneath PHA, Sokal RR (1973) Numerical taxonomy. The principles and practice of numerical classification. WH Freeman and Co, San Francisco, pp 573

Vladik P, Vitovec J, Cervinka S (1974) The taxonomy of Gram-positive immobile diplobacilli isolated from necrotizing nephroses in American char and rainbow trout. Vet Med 19:233–238

Whittenbury R (1964) Hydrogen-peroxide formation and catalase activity in the lactic acid bacteria. J Gen Microbiol 35:13–26

Infection with an Acinetobacter-like Bacterium in Atlantic Salmon (Salmo salar) Broodfish

S.O. ROALD and T. HASTEIN

Sixty sexually mature Atlantic salmon *(Salmo salar)* ranging in weight from 5–12 kg were collected in the autumn of 1978 from the wild population in the river Surna and transported to a holding facility in brackish water, where they were kept for breeding.

There was an outbreak of disease in this group of fish, characterized by ulceration of the skin at the bases of the pelvic and anal fins. Mortality among the clinically affected fish began at the end of September and lasted for about five weeks, when water temperatures ranged from 8°–11 °C. About 40% of the fishes showed clinical signs, and the case fatality rate was 55.

Four diseased fish were examined for pathological lesions and a microbiological examination was made. In the skin at the bases of affected fins, the earliest changes comprised hyperemia of dermal vessels and hemorrhages in the scale pockets with severe edema extending into the lower epidermis. Ulceration followed quickly and the lesions extended down into the underlying muscle tissue, where myofibrillar dry necroses, hemorrhages, and general cellular inflammatory infiltrations were found. Focal necroses appeared in the liver, spleen, and kidney. Multiple small hemorrhages were found in the swim bladder and on the viceral peritoneal surfaces.

Microscopical examination of smears from the necrotic foci and internal organs showed the presence of numerous nonmotile gram-negative rods. The bacteria were isolated in pure cultures from heart blood, liver, kidney, spleen, and from the skin ulcers. After incubation at 22 °C for 48 h the microbe appeared as a short and plump gram-negative rod measuring $0.8-1.2\ \mu \times 1.6-1.8\ \mu$. It was nonmotile and not acid-fast. No spores were formed.

On 5% blood agar plus 0.5 sodium chloride the colonies were round, mucoid, raised, translucent, and measured approximately 1.5 mm in diameter after 48 h, incubated aerobically. A distinct zone of hemolysis was present, which increased on further incubation. The bacterium was facultative anaerobic. The biochemical properties of the microbe are shown in Table 1.

Investigations (genetic studies and examinations by gas chromatography) made at the Kaptein W. Wilhelmsen og Frues Bakteriologiske Institutt, Oslo (K. Børve, personal communication) on the isolated bacterium indicated that this psychrophilic bacterium belongs to the family *Neisseriacea* and may be related to *Acinetobacter* or to *Moraxella*, or may represent an intermediate group between the two genera. Further taxonomic studies of this organism are needed.

---

1 National Veterinary Institute, Postbox 8156, Dept., Oslo 1, Norway

Table 1. Characteristics of the *Acinetobacter*-like isolate

Test	Results	Test	Results
ONPG test	−	Motility	−
Arginine dihydrolase	−	Adonitol	−
Lysine decarboxylase	−	Maltose	+
Ornithine decarboxylase	−	Salicin	−
Simmons citrate	−	Lactose	−
Production of H_2S	−	Raffinose	−
Urease (Ferguson)	−	Dextrin	−
Tryptophane desaminase	−	Inulin	−
Indole	−	Monnose	+
Acetoine	−	Galactose	+
Proteolysis of gelatin	−	Cellobiose	−
Glucose	−	Trehalose	−
Mannitol	−	Oxidation	−
Inostitol	−	Fermentation	−
Sorbitol	−	MacConkey	−
Rhamnose	−	Nitrate reduction	−
Saccharose	−	Production of nitrogen	−
Melibiose	−	Chloramphenicol	+
Amygdaline	−	Penicillin (high)	+
L (+) arabinose	−	Erythromycin	+
Cytochrome oxidase	+	Neomycin	+
Dimethyl oxidase	−	Tetracyklin	+
Tetramethyl oxidase	+	Novobiocin	+
Catalase	+	Sulfa	+
Voges Proskauer	−	Vibriostat	−
Methyl-red	−		

Pathogenicity tests with peptone water cultures inoculated on 15-g salmon fingerlings resulted in the death of the fishes within 72 h when the organism was administered by the intramuscular route at a water temperature of 12 °C.

The disease outbreak was controlled by treatment with antibiotics. In this case a single intramuscular dose of 100 mg oxytetracycline chloride per kg fish gave therapeutic effect.

Critical problems in the study of bacterial pathogens of fish are the correct identification of the infectious agent and the judgment of its role as primary or secondary invader. Many of the bacteria normally present in sea water or on the surface of fish can invade and cause pathological effects if fish are injured or subjected to other severe environmental stress. *Acinetobacter* has been isolated from the digestive tract of salmonids in freshwater and sea water (Shewan, 1961; Trust and Sparrow, 1974; Roald, 1977). *Acinetobacter* and related bacteria also belong to the principal genera apprearing on the skin and gills of Atlantic salmon from coastal, estuarine, and river waters (Horsley, 1973). In this case the organism seemed to be highly pathogenic to the mature salmons by natural infection and to the inoculated salmon fingerlings from which the bacterium could be reisolated. The described microbe is therefore considered a definite pathogen.

Acknowledgment. The authors wish to thank Dr. K. Bøvre of the Kaptein W. Wilhelmsen og Frues Bakteriologiske Institutt, Oslo, for his assistance with the genetic studies and the examinations by gas chromatography.

References

Horsly RW (1973) The bacterial flora of the Atlantic salmon *(Salmo salar L.)* in relation to its environment. J Appl Bacteriol 36:377–386

Roald SO (1977) Effects of sublethal concentrations of lignosulphonates on growth, intestinal flora and some digestive enzymes of rainbow trout *(Salmo gairdneri)*. Aquaculture 12:327–335

Shewan JM (1961) The microbiology of sea-water fish. In: Bergstöm G (ed) Fish as food, vol I. Academic Press, London New York, pp 487–560

Trust TJ, Sparrow RAH (1974) The bacterial flora in the alimentary tract of fresh-water salmonid fishes. Can J Microbiol 20:1219–1228

"Sphaerosporosis", a New Kidney Disease of the Common Carp

K. MOLNÁR[1]

Parasitological surveys of cultured one-summer carp often revealed the presence of spores and developing stages of the myxosporidium *Sphaerospora angulata* Fujita, 1912 in renal tubules (Molnár, 1980). The rate of *S. angulata* invasion was especially high in fish stocks also affected by other parasitic or nonparasitic diseases (blood protozoans, gill sphaerosporosis, swim bladder disease). This prompted closer investigations into carp renal sphaerosporosis, a disease not previously known in Europe. The study covered the following aspects: (1) nature of renal lesions caused by *S. angulata;* (2) possible identity of the extracellular blood sporozoan of unknown systematic position described by Csaba (1976) with a developmental stage of *S. angulata;* (3) etiological responsibility of one or two different *Sphaerospora* spp. for the gill and kidney sphaerosporosis of carp; (4) possible causal involvement of *S. angulata* in the etiologically unknown swim bladder disease associated with swelling of the kidney.

Of the many known *Sphaerospora* spp. only the histozoic species *S. carassii, S. tincae,* and *S. reichenowi* have hitherto been regarded as pathogenic (Leger, 1930; Plehn, 1932; Hámory and Molnár, 1972; Kashkovskij et al., 1974; Molnár, 1979; Jacob, 1953). Information has been scarce on the pathogenicity of coelozoic parasites, of which little is known apart from their prevalence. In the USSR 15 such species have been identified (Shulman, 1966).

Of the *Sphaerospora* spp. parasitic in carp, only *S. carassii* has been shown to infect indigenous European hosts (Hámory and Molnár, 1972; Lom et al., 1976; Molnár, 1979), while *S. cyprini* and *S. angulata* were originally found only in Far Eastern habitats (Fujita, 1912; Shulman, 1966). Recently Razmashkin and Skriptsenko (1976) reported the occurrence of *S. cyprini* in Western Siberia, and Osmanov (1971) that of *S. angulata* in Central Asia.

Certain authors have taken into consideration the etiological involvement of protozoa in swim bladder disease. Since this condition is associated with renal hypertrophy, and protozoon-like bodies are often found in the wall of the affected air bladder (Szakolczai, 1967), the causal importance of sphaerospores cannot be excluded with certainty.

Observations on the occurrence, development, etiology, and pathogenesis of renal sphaerosporosis, and on its associations with infection by the blood sporozoan of Csaba (1976), gill sphaerosporosis, and swim bladder disease, are reported in this paper.

---

[1] Veterinary Medical Research Institute, Hungarian Academy of Sciences, Budapest, Hungary

Material and Methods

Carp populations from nine Hungarian pond farms were regularly screened for renal sphaerosporosis from 1976 to 1978. For the most one- and two-summer carp were examined, but occasionally older carp and other fish species from the canal system of the farms were also included in the study. During 1977–1978 fry and one-summer hosts, 20 of each from November 1977 to May 1978, were taken at biweekly intervals for parasitological examination from the two most heavily infected farms, in which fry rearing was also carried out. Older carp and other fish species were occasionally examined along with the regular sample.

Impression smears of the visceral organs, especially the kidneys, were examined microscopically and two infected kidneys from each sample were examined histologically.

In 1978 and 1979 complementary studies were performed on the relationship of renal sphaerosporosis with concomitant diseases. These studies covered the parasitological examination of 80 (one- to three-month-old, 156 three-to six-month-old carp fry and 32 overwintered one-summer and two-summer carp. Gross examination was followed by the study of impression smears from the blood, kidneys, gills, and swim bladder.

Six young carp reared in the laboratory under parasite-free conditions were infected experimentally by the intra-abdominal route with blood from a fish harboring Csaba's protozoa (C-hemoprotozoa). Gill biopsies were taken regularly and examined microscopically for sphaerospores, and two fish of each group were killed at appropriate intervals to examine the kidneys for renal sphaerosporosis.

For histological examination the organ specimens were fixed in 10% formalin or Bouin's solution, embedded in paraffin wax, and stained with hematoxylin and eosin, Farkas-Mallory technique or Gömöri's trichrome stain.

Results

S. angulata spores and developmental stages were detected in the kidney of several hosts in each farm under survey, but the percentage occurrence of the parasite differed between farms, ranging from 10%–15% to more than 50%.

The kidneys of hosts with sphaerosporosis showed no gross changes compared to noninfected controls.

The *S. angulata* spores and developmental stages equally localized in the renal tubules. Both could be easily identified in the impression smears. The mature spores had a characteristic angular shape, while the pansporoblasts and young spores had a granular appearance. The latter was due to the presence of 8–14 conspicuous nuclei in the 12–15-μm-wide pansporoblasts; 6 nuclei were still present in the developing spores.

Renal sphaerosporosis has usually been demonstrated at 1.5–3 months of age in the fry at the earliest, but quite recently we also detected the developing stages of *Sphaerospora* in 1.5-month-old hosts. The intensity of infection reached a peak at 4–6 months of age. Among the older fish infected individuals were found in all

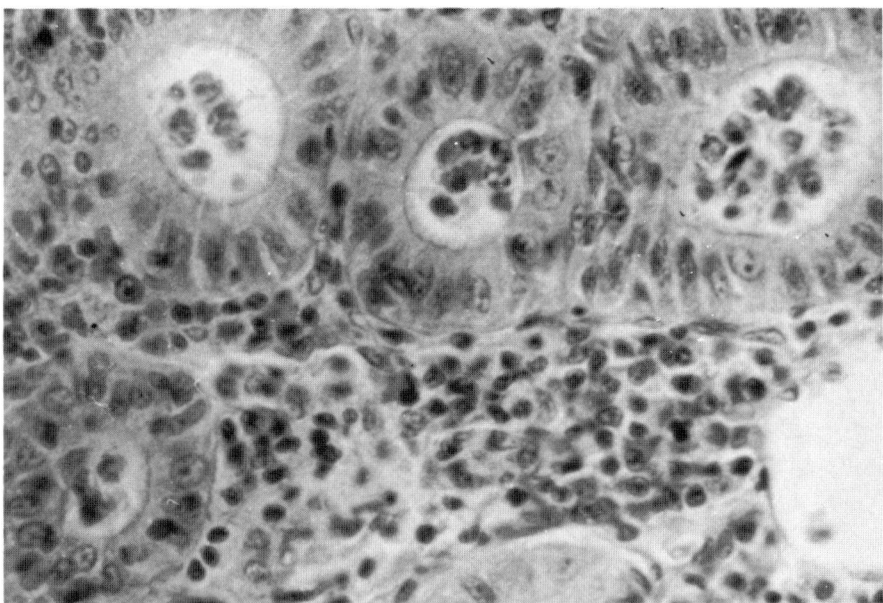

Fig. 1. Tubular lumen packed with sphaerospores. HE, x 300

seasons of the year, but the number of hosts harboring spores was low among the two-summer population.

Confirming the findings from unstained impression smears, histological examination also showed that the parasites localize exclusively in the tubular lumen, without invading the lining epithelium or parenchyma. Occasionally 80% of the tubules were filled with *S. angulata* stages (Fig. 1). In mild infections the parasites established themselves chiefly in the proximal segment of the tubules, whilst in massive infections they also appeared in the intermediate and distal segments.

The earliest stages of *S. angulata* found in renal tubules were multinucleated pansporoblasts; these contained developing pairs of spores at a more advanced stage of the infection (Fig. 2). Spore development was not synchronous. While some segments harbored mature spores, others were filled with developing forms. Furthermore, in some tubules the mature spores were surrounded by developmental stages. (Mature spores take on a yellow, developing spores a blue or red Farkas-Mallory stain.) In some preparations reticular residues of pansporoblast, with trapped spores, indicated an earlier infection. In the same preparations the narrow lumen of some distal tubular segments contained, in addition to spores, condensed tissue debris and starlet-shaped, needle-shaped, or cylindrical salt crystals. Presumably the condensed mass assumed a vivid red colour on staining with hematoxylin and eosin; it filled the lumen of the tubules as a compact cast (Fig. 3) and accounted through obliteration for a functional insufficiency of the nephrons.

Light-microscopic examination revealed no interaction between the tubular lining epithelium and the mass of parasites filling the lumen. However massive the infection, the tubular epithelium and its brush border appeared intact, although slightly reduced

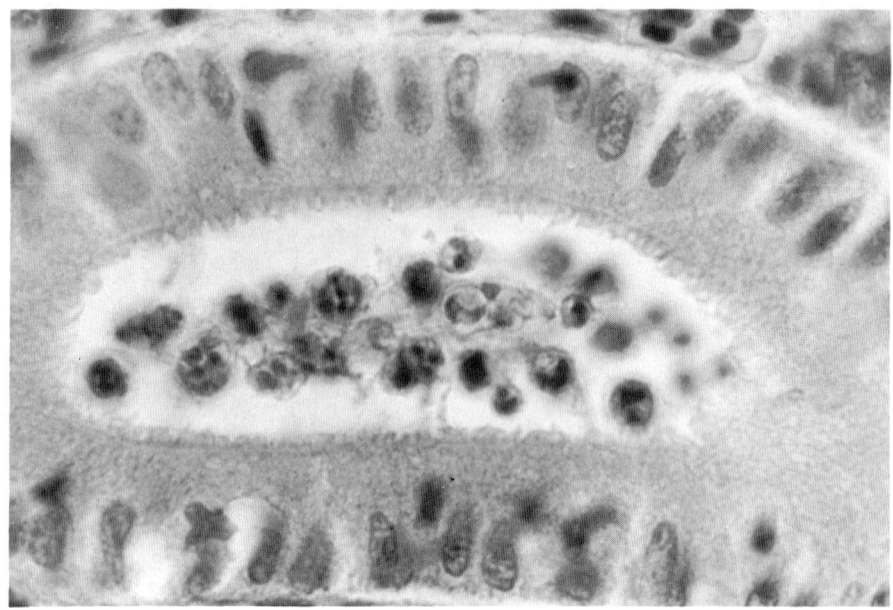

Fig. 2. Spores and developmental stages in a renal tubule. HE, x 500

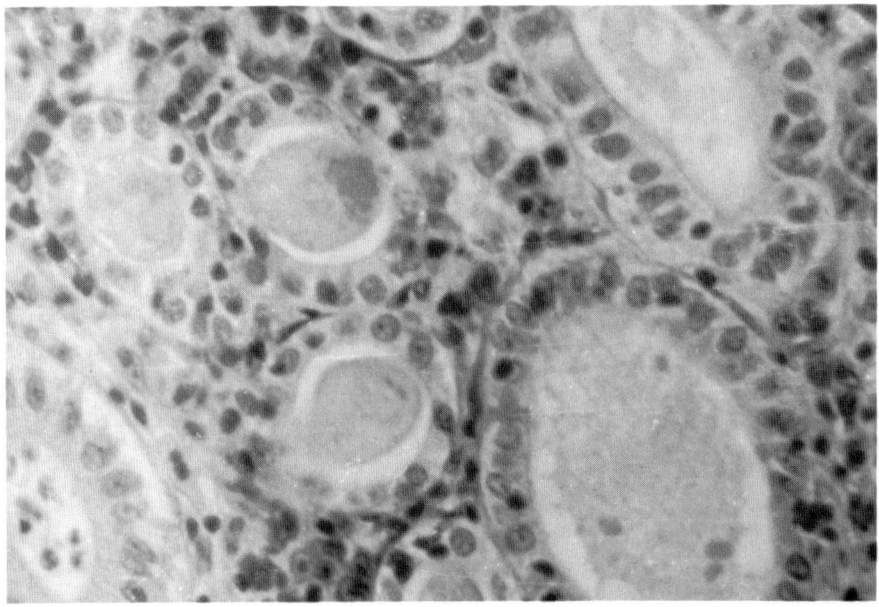

Fig. 3. Accumulation of tissue and other casts in renal tubules in consequence of *S. angulata* infection. HE, x 300

in height owing to dilation of the parasite-packed lumen. Epithelial atrophy followed only upon obstruction of the tubular lumen.

Among the 80 one- to three-month-old carp fry used for examination of the involvement of *S. angulata* in mixed infections only two harbored *S. angulata* in the kidney. Spores of the gill parasite *S. carassii* were found in 16 cases, and 9 hosts harbored the C-hemoprotozoon. No host had swim bladder disease. No simultaneous occurrence of the two *Sphaerospora* species, or of either together with C-hemoprotozoon was observed in this age group.

Among the 156 three- to six-month-old hosts 36 (23%) had been infected by *S. angulata*. *S. carassii* infection did not occur in this group. C-hemoprotozoa were found in 33 cases (21.2%), and 17 fish (10.9%) had swim bladder disease.

The analysis of mixed infections has shown that the air bladder disease occurred in association with *S. angulata* and C-hemoprotozoon infection in 4 cases (2.6%), with *S. angulata* infection alone in 12 cases (7.6%). A mixed C-hemoprotozoon and *S. angulata* infection was found in 17 hosts (10.9%). Among 17 hosts affected by air bladder disease 12 (70.5%) and among 33 hosts with C-hemoprotozoon infection 17 (51.1%) were simultaneously diseased with renal sphaerosporosis. Among the 139 hosts free from the swim bladder disease 24 (17.3%) had *S. angulata* infection.

Among 32 overwintered one-summer (6 months old or older) fish 5 harbored *S. angulata* and of these 3 also had swim bladder disease. The C-hemoprotozoon was found in 7 hosts, of which one also harbored *Sphaerospora*.

The spores isolated from the gills differed from those found in the kidneys in both shape and size. Spores from the gills were 9–12 μm in diameter, spherical in shape, with a smooth surface, and the two halves of the spore wall were connected by a thick, prominent suture. The polar capsules were pyriform, 4–5 μm long by 3–4 μm wide. The spores found in the renal tubules were not round but roundish, slightly tapering at one end (at the polar capsule), but somewhat angular at the other, with tiny ear-like projections on their surface. The renal spores were 6–7.5 μm long by 6–6.5 μm wide. The two polar capsules were nearly equal in size, 3.5–4 μm long by 2–2.5 μm wide. The suture uniting the two halves of the spore wall rose only slightly above the surface. Morphologically the spores found in the gills corresponded to *Sphaerospora carassii* Kudo, 1919, those found in the kidney to *S. angulata* Fujita, 1912.

As well as in blood smears the C-hemoprotozoon was detected in impression smears and histological sections from kidneys, gills, and swim bladder, but always within the circulation system (Fig. 4). C-hemoprotozoa were especially numerous in gill vessels with narrow lumina. The microscopic appearance of the 6–7-μm-long parasites, with their 6 or 8 nuclei, was reminiscent of the pansporoblast stages of *S. angulata* and *S. carassii*. Susceptible hosts inoculated intra-abdominally with blood containing C-hemoprotozoa always became infected by these parasites. The intensity of infection tended to increase until the end of the first month, then it gradually became stabilized at a lower level over the next two months. The hosts killed 1, 2, and 3 months after experimental infection did not harbour sphaerospores in the gills or in the kidney.

We failed to detect the parasite-like or fungus-like bodies described by Szakolczai (1967) in impression smears of the changed swim bladder, but we learned from Kovács-Gáyer by personal communication that similar formations have been frequently seen in histological sections from affected swim bladder.

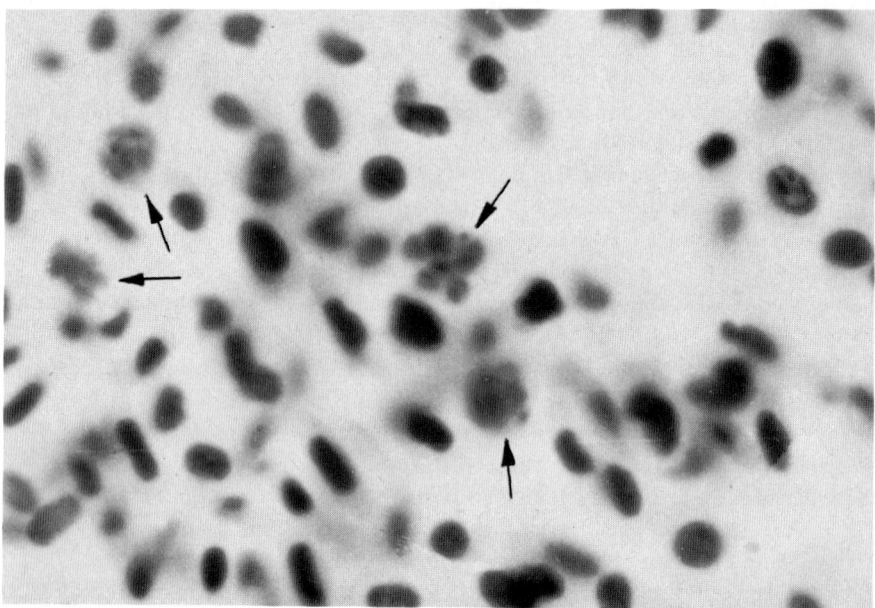

Fig. 4. Csaba's blood protozoon *(arrows)* in the gill vein. HE, x 800

Discussion

The present studies have shown that renal sphaerosporosis is a fairly frequent parasitosis in carp 3 months old and older in Hungary. Since gill sphaerosporosis was not infrequently also present in the affected host populations, it was postulated that the two conditions might possibly be due to stages of the same species established in different locations depending on the time of infection. The fact that not a single host was simultaneously diseased with gill and renal sphaerosporosis weighed in favor of this hypothesis, but the dissimilar morphology and size of the spores found in gill and kidney unequivocally indicated that they represented independent species, *S. carassii* and *S. angulata* respectively.

The spores and developing stages of *S. angulata* which filled the renal tubules in large masses are obviously coelozoic parasites and as such do not directly damage the tubular epithelium. But despite their ceolozoic nature they definitely affect the host organism by utilizing nutrients excreted but not yet reabsorbed by the kidney and, especially, by mechanical obstruction of the renal tubules. The salt crystals precipitating from retained urine, and debris resulting from spore development form in the obliterated distal tubules a homogeneous conglomeration which might cause nephrosis.

Leger (1930) and Plehn (1932) similarly established that the renal sphaerospora *S. tincae,* a parasite of the tench, accounted for a considerable enlargement of the kidney. With the renal *S. angulata* infection we saw no gross renal hypertrophy except in hosts having swim bladder disease as well. We have therefore regarded the enlargement of the kidney as a sequel to the latter condition.

Among the Myxosporidia parasitic in the renal tubules of the carp, only the species *Hoferellus cyprini,* morphologically very different from *S. angulata,* has hitherto been known in Europe. The spores and pansporoblasts of *H. cyprini* also localize in the renal tubules, but Plehn (1924) detected even its trophozoites in the tubular wall. In Hungary we failed to detect *H. cyprini* in the large host material covered. The hitherto unkonwn early trophozoites of *S. angulata* probably develop outside the renal tubules. This circumstance suggests the identity of *S. angulata* with the C-hemoprotozoon, particularly if it is taken into consideration that 51.5% of the hosts infected with the latter also harbored *S. angulata.* On the basis of light-microscopic morphology their identity cannot be excluded. A proof against identity is, however, presented by the fact that neither hosts with spontaneous C-hemoprotozoon infection, nor susceptible hosts experimentally infected with C-hemoprotozoon containing blood developed sphaerosporosis during a long period of observation in the laboratory. The frequent simultaneous occurrence of the two species is in all probability due to inadequate environmental conditions which favor the invasion of both.

A closer relationship seemed to exist between renal *S. angulata* invasion and swim bladder disease. Among fish with swim bladder disease 70.5% also had renal sphaerosporosis; this represents a four fold greater incidence than that found in unaffected populations. It seems unlikely that *S. angulata* giving rise to only minor renal damage, could be the direct cause of the swim bladder disease, but the parasite-like bodies observed by Kovács-Gáyer in the bladder wall need further detailed studies. It is quite likely that a decreased resistance may play the major role in the relationship of the two diseases, i.e., fish with decreased resistance are more susceptible to secondary infections or invasions. On the basis of the present findings it cannot be decided with certainty whether swim bladder disease or *S. angulata* invasion is primarily or secondarily involved.

References

Csaba GY (1976) An unidentifiable extracellular sporozoan parasite from the blood of the carp. Parasitol Hung 9:21–24

Fujita T (1912) Notes on new sporozoan parasites of fishes. Zool Anz 39:259–262

Hámory GY, Molnár K (1972) A protozoan disease of the fry in fish farms. Magy Allatorv Lapja 27:358–360

Jacob E (1953) Eine bislang unbekannte Sphaerosporose des Flußaals hervorgerufen durch *Sphaerospora reichenovi* nova species, mit eigenartigem Sitz im Darm. Berl Münch Tierärztl Wochenschr 66:326–328

Kashkovskij VV, Razmashkin DA, Skriptsenko EG (1974) Diseases and parasites of fish farms in Siberia and Ural. Sredne-Uralskoe Knizhnoe Izdatelstvo Sverdlovsk (In Russian) pp 1–159

Leger L (1930) Une nouvelle maladie parasitaire funeste aux elevages de tanche "la Sphaerosporose". Trav Lab Hydrobiol Piscicult Uni Grenoble 21:7–13

Lom J, Golemansky V, Grupcheva G (1976) Protozoan parasites of carp *(Cyprinus carpio L.)*: A comparative study of their occurrence in Bulgaria and Czechoslovakia, with description of *Trichodina perforata* sp. n. Folia Parasitol, Prague 23:289–300

Molnár K (1979) Gill Sphaerosporosis in the common carp and grass carp. Acta Vet Acad Sci Hung 27:99–113

Molnár K (1980) Renal Sphaerosporosis in the common carp *Cyprinus carpio L.* J Fish Dis 3:11–19

Osmanov SC (1971) Parasites of fish in Uzbekistan. Izdatelstvo FAN Uzbekskoj SSR. Taskent. (In Russian) pp 1–532

Plehn M (1924) Praktikum der Fischkrankheiten. Handb Binnenfisch Mitteleur 1:301–470
Plehn M (1932) Eine Schleienbrutkrankheit und ihr Erreger *Sphaerospora tincae* n sp. Int Rev Gesamten Hydrobiol Hydrogeogr 26:265–280
Razmashkin DA, Skriptsenko EG (1976) Diseases of fish in the pond farms of Siberia and Ural. Tiumen, pp 89–97 (In Russian)
Shulman SS (1966) Myxosporidia in the faune of the USSR. Izd Nauka, Moscow-Leningrad (In Russian), pp 1–507
Szakolczai J (1967) Untersuchungen der Schwimmblasenentzündung bei Karpfen anhand von zwei Fällen in Ungarn. Z Fisch Hilfswiss 15:139–151

Disease in Farmed Juvenile Atlantic Salmon Caused by Dermocystidium SP.

A.H. McVICAR and R. WOOTTEN[1]

Introduction

Dermocystidium has been recorded from oysters (Mackin et al., 1950), numerous species of freshwater fish and some Amphibia (Reichenbach-Klinke and Elkan, 1965). The taxonomic affinities of the group have been uncertain until recently when Perkins (1976) demonstrated that *Dermocystidium marinum* from oysters, previously considered by many workers to be a fungus is, in fact, a member of the protozoan phylum Apicomplexa, since it possesses a zoospore stage containing an apical complex. Subsequently Levine (1978) erected a new genus *Perkinsus* to distinguish the oyster parasite from those found in fish and Amphibia and placed it in a new class Perkinsea, within the Apicomplexa. The relationship between the invertebrate and vertebrate parasites is not clear because of the lack of information on developmental stages of the latter group.

Those *Dermocystidium* occurring in fish and Amphibia usually form subcutaneous or gill cysts (Reichenbach-Klinke and Elkan, 1965). Severe gill pathology and mortalities caused by *Dermocystidium* have been reported in pre-spawning adults and emergent fry of chinook salmon *(Oncorhynchus tschawytscha)* in the USA (Pauley, 1967; Allen et al., 1968).

Epidemiology

A visceral form of *Dermocystidium* was found in parr of Atlantic salmon *Salmo salar* L. from a fish farm in north-west Scotland in winter 1977–1978. Fish were reared in tanks supplied with water drawn unfiltered from a river. Up to 3% of some populations were affected in December 1977. Parr were held at densities of about 3,000 fish per 1-m^2 tank. Fish with gross signs of disease had first been observed in October 1977. Only fish in their first year of life were apparently affected.

A further sampling of the farm in February 1978 using in vitro incubation of visceral fat revealed only one infected fish out of 75 parr taken from three populations. Clinically diseased fish were very scarce on the farm at this time and have not been observed subsequently. An examination of 100 parr from four populations using in vitro incubation (See Sect. "In vitro Culture") in October 1978 did not detect the parasite.

---

[1] DAFS, Marine Laboratory, Victoria Road, Aberdeen, Scotland

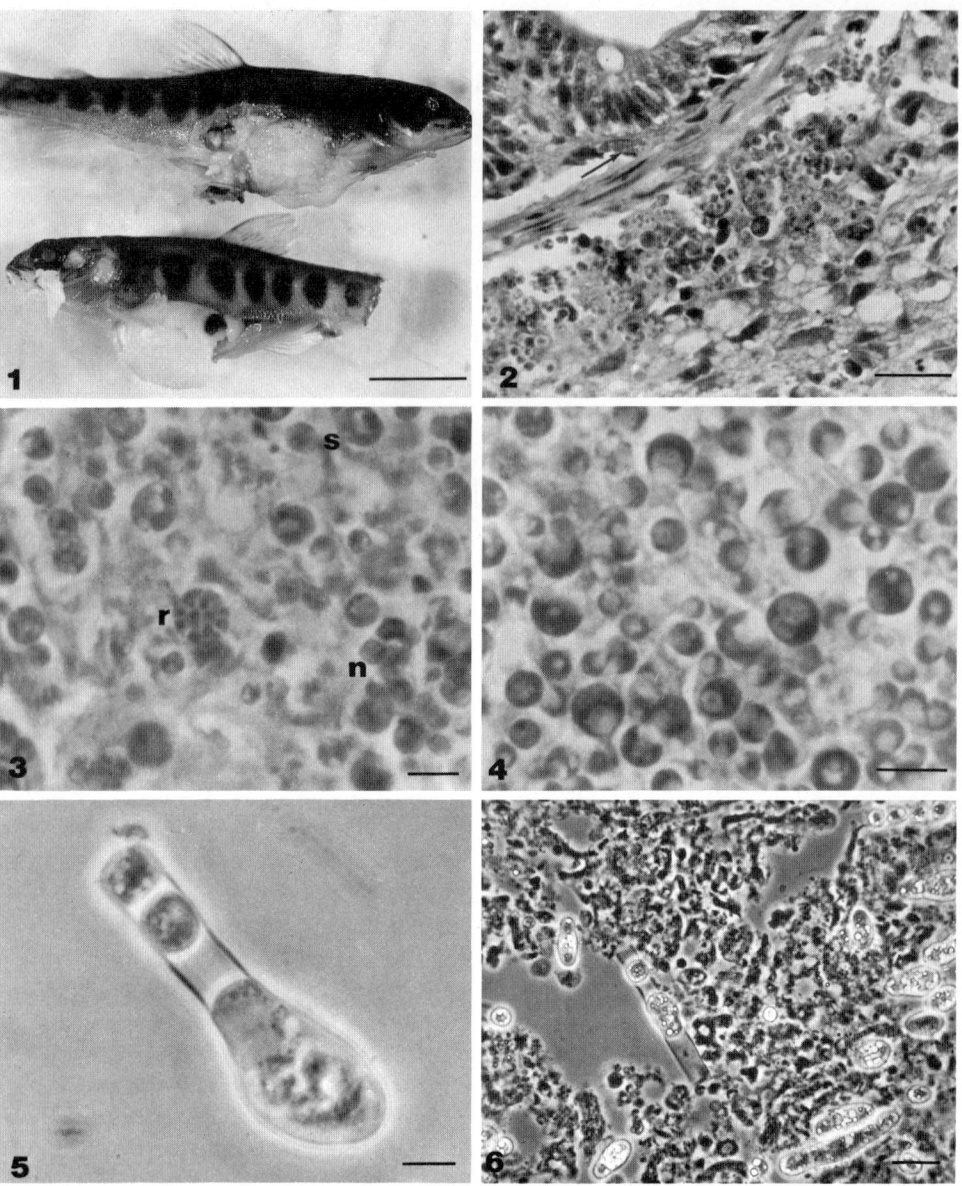

Fig. 1. Atlantic salmon parr with latero-ventral abdominal wall removed to show the caseous *Dermocystidium* lesion. *Bar*, 1.0 cm

Fig. 2. *Dermocystidium* lesion surrounding a pyloric caecum indicating minimal invasion of the gut by the parasite *(arrowed)*. *Bar*, 25 μm

Fig. 3. *Dermocystidium* cells situated towards the periphery of a nodule showing non-vacuolated cells *(n)*, rosette *(r)*, and signet-ring *(s)* stages. *Bar*, 10 μm

Fig. 4. *Dermocystidium* cells situated towards the centre of a nodule where signet-ring cells predominated. *Bar*, 10 μm

Fig. 5. In vitro development of *Dermocystidium* in thioglycollate medium. Phase contrast illumination. *Bar*, 10 μm

Fig. 6. A positive screening test for *Dermocystidium* in salmon visceral fat after incubation in thioglycollate medium. *Bar*, 25 μm

A total of 175 parr examined from two other widely separated fish farms in Scotland in February, October, and December 1978 were also found to be uninfected using in-vitro incubation.

Pathology

Affected salmon were obvious because of their swollen abdomen and abnormal behaviour. Such fish remained motionless on the tank bottom or swam only with difficulty.

Dissection of affected fish typically revealed an extensive yellow-white caseous mass in the visceral cavity, particularly associated with the fat body around the pyloric caeca and spleen (Fig. 1). Although the viscera were often deeply embedded in this growth they were easily detached from it, retaining their normal appearance and integrity. There was no evidence of extensive replacement of organ tissue by parasite cells. Constriction of viscera, especially the intestine, was inevitable due to the size of the lesion in advanced infections and this, together with eventual rupture of the abdominal wall, was the cause of death in affected fish.

Tissue for light microscopy was fixed in 10% phosphate-buffered formol saline and embedded in paraffin wax. Sections were cut at 5 μm and stained with haematoxylin and eosin (HE). Material for electron microscopy was fixed in 2.5% glutaraldehyde in Millonig's buffer with a post-fixation in 1% osmium tetroxide. After dehydration in graded alcohols the material was embedded in Epon. Ultrathin sections were stained with uranyl acetate and lead citrate and examined in an AEI EM 6G electron microscope.

Histologically the growth consisted largely of parasite cells arranged into a series of nodules or foci of infection, up to 1.2 mm in diameter. The centres of these foci were made up of necrotic tissue with only occasional parasite cells, but the latter increased in number towards the outside of the foci so that at the periphery the tissue was almost totally parasitic (Figs. 2–4).

There was no definite limiting capsule surrounding each nodule or any other marked host response restricting spread of the infection. In the outermost parts of nodules and between nodules groups of parasite cells or single cells were enmeshed in a fibrous network (Fig. 2) which, when viewed with the electron microscope (Fig. 9), could be seen to consist of collagen bundles, host cell debris and occasional macrophages. There was no extensive cellular infiltration of nodules. Although the evidence suggests a development of the parasite outwards from foci of infection the heart and gills were the only organs to show accumulations of parasite cells and only in the latter were macroscopic nodules occasionally visible. *Dermocystidium* cells were also observed in small numbers in liver, spleen, and kidney sections. In such cases there was no apparent host response, the parasites usually being lodged in host capillaries. There was occasional invasion of the peripheral tissue of liver, spleen, and gut from adjacent parasite nodules.

Parasite Morphology

Dermocystidium cells were small (2.5–8.0 μm), spherical and densely staining with HE. Several forms in a developmental sequence could be distinguished with the light and electron microscope. The most common cells had large spherical nuclei centrally situated with a prominent electron-dense endosome (Figs. 3 and 7). These cells underwent multiplication by division of the nuclei (to form multinuclear cells; Fig. 7) and cleavage of the cytoplasm leading to the formation of daughter cells by infolding of the cell wall membranes (Figs. 8 and 9). The daughter cells rounded up to form rosette-like bodies and were finally liberated by rupture of the original cell wall (Figs. 3 and 7). Nuclei in the process of division were not observed. The cytoplasm of the cells was densely granular and contained occasional mitochondria, numerous membrane-bound vesicles and extensive membranes which were not organised into recognisable organelles. The cell walls consisted of a typical trilaminate membrane overlaid by a granular fibrillar matrix which formed during early cleavage of the mother cell and was not therefore of host origin (Fig. 8). The outer matrix persisted even after rupture of a cell and release of daughter cells.

A probable developmental sequence could also be traced from the actively dividing cells described above to a cell type characterised by a large cytoplasmic vacuole, the signet-ring stage (Figs. 4 and 10). The vacuole appeared to develop through the coalescence of cytoplasmic vesicles and it typically grew to occupy most of the cell, so that the cytoplasm was restricted to a narrow peripheral border. A discrete nucleus was not usually observed in sections of typical signet-ring stages. Membrane-bound bodies and vesicles with varying electron opacity were often observed in the vacuoles and complex structures consisting of convoluted membranes, vesicles and tubules (vacuoplasts) were usually associated with, or contained within, the vacuoles. These complexes were also associated with the cell wall membranes, both in vacuolated and non-vacuolated cells. In the latter they possibly represent lomasomes. Signet-ring stages were most common towards the centre of parasite nodules, while rosettes and non-vacuolated cells predominated towards the periphery.

In Vitro Culture

In vitro development of *Dermocystidium* was achieved by aseptic removal of small pieces of visceral lesions from infected salmon, followed by vigorous agitation to break up the tissues and culture under aerobic conditions at constant temperatures in a variety of media. Uncontaminated cultures were regularly monitored by subsampling and determining cell density, on an improved Neubauer haemocytometer. In the fluid thioglycollate medium tabulated by Perkins (1966) little increase in parasite cell numbers was observed but in both Oxoid thioglycollate in distilled water and minimum Eagle's medium +1% foetal calf serum (MEM) multiplication was rapid.

Cells in MEM were typically small and derived from an internal cleavage and rupture of spherical mother cells, whereas those in Oxoid thioglycollate were larger and normally formed by progressive cleavage of elongate cells which often formed a tube through which the daughter cells passed (Fig. 5). Sporulation did not occur in culture and could not be induced by transfer of cells to sterile sea water or freshwater. Perkins (1974) also failed to induce zoosporulation in *Dermocystidium* from salmon.

Using Oxoid thioglycollate medium optimum temperature for growth was 10 °C with an approximately tenfold increase in cell density within four days (Fig. 11). The sigmoid growth pattern apparent around the optimum temperature was associated

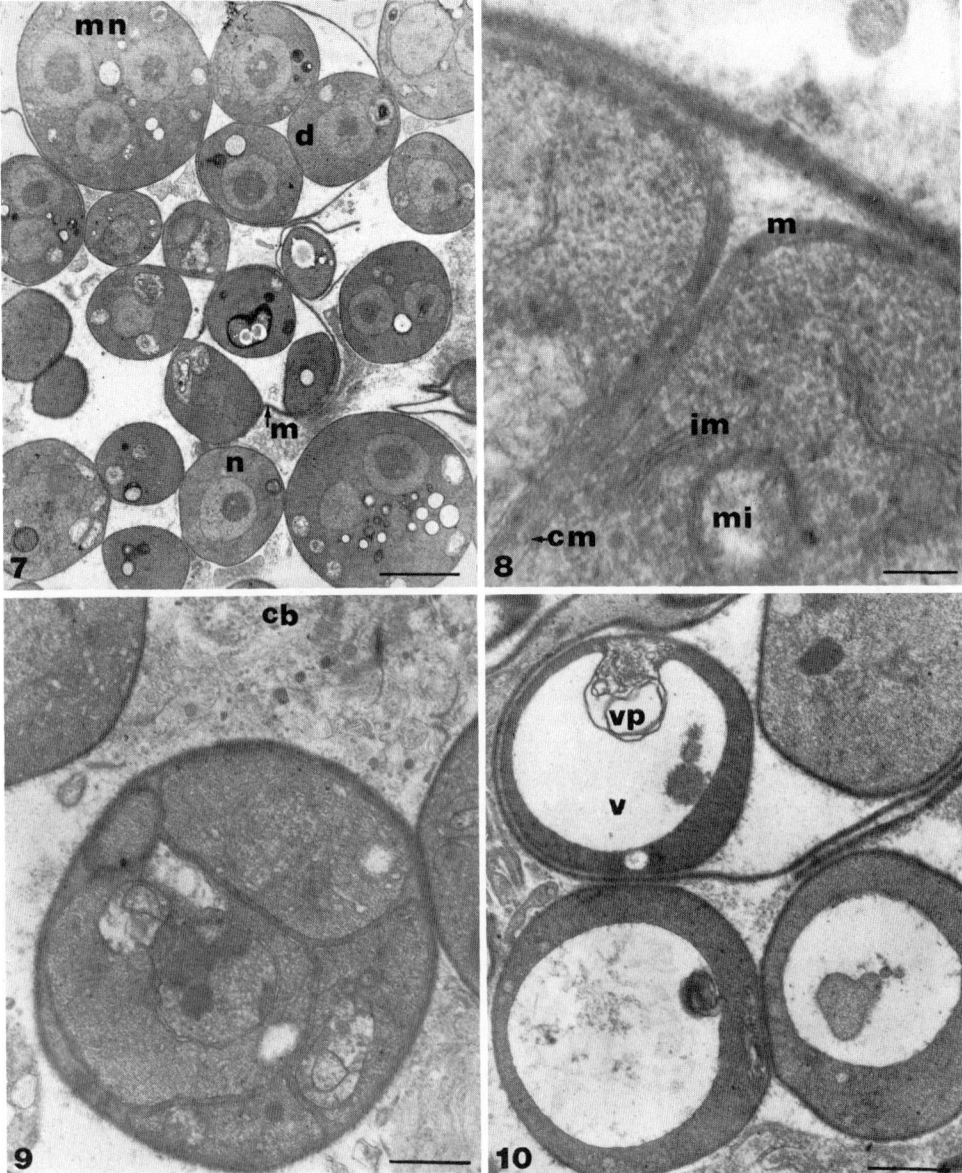

Fig. 7. Non-vacuolated cells *(n)*, multinucleate cells *(mn)*, rounded-up daughter cells *(d)*, and liberation of daughter cells after rupture of the mother cell wall matrix *(m)*. *Bar,* 5 μm

Fig. 8. Detail of the formation of daughter cell walls beneath the matrix of the mother cell wall. Note the trilaminate cell membrane *(cm)* overlaid by the granular fibrillar matrix *(m)*, the granulation of the cytoplasm and the presence of intracytoplasmic membranes *(im)*, and a mitochondrion *(mi)*. *Bar,* 0.2 μm

Fig. 9. *Dermocystidium* cell showing cleavage of cytoplasm before rounding-up of daughter cells. Collagen bundles *(cb)* of a fibrous host cell are illustrated. *Bar,* 1 μm

Fig. 10. Signet-ring cells, showing the large vacuole *(v)*, vacuoplast *(vp)* and other vacuolar inclusions. *Bar,* 1 μm

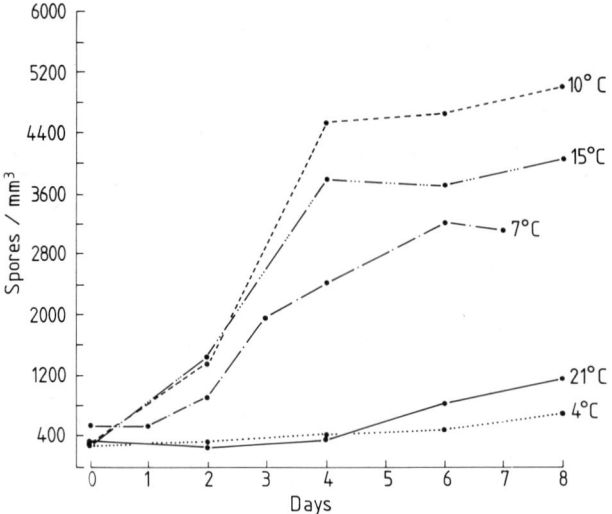

Fig. 11. Growth of *Dermocystidium* sp. from Atlantic salmon in thioglycollate medium at various temperatures

with depletion of essential nutrients as cultures in which growth had slowed could be revitalised by adding new medium.

The extensive in vitro multiplication was successfully utilised as a basis of a screening test for *Dermocystidium* in salmon without the necessity of staining with Lugol's iodine as recommended by Ray (1952) for diagnosis of *D. marinum* in oysters. In our tests small pieces of visceral fat tissue were incubated in Oxoid thioglycollate at 10 °C for 3–5 days. On examination cultures from infected fish showed massive development of the parasite as described above (Fig. 6).

Transmission

Attempts were made by various means to transmit *Dermocystidium* between fish. Salmon fry were successfully infected by injecting a lesion homogenate in Hank's balanced saline prepared from a naturally infected fish into the stomach perorally and by immersing fish for 3 h in a suspension of a lesion homogenate. Fish were maintained at 10 °C and were tested for infection by in vitro incubation of fat tissue. Five out of six fish given a peroral injection of lesion were found to be infected when tested after 46–83 days. One of the fish showed a gross lesion in the visceral fat tissue. Only one of the six salmon immersed in a lesion homogenate became infected. Fish from this experiment were tested 38–101 days after exposure.

Salmon fry exposed to effluent water from naturally infected fish, or fed food sprayed with a lesion homogenate, did not become infected. Three out of six yearling rainbow trout given an intraperitoneal injection of 0.8–1.0 ml of lesion homogenate in Hank's balanced saline were found to have subclinical infections when examined 35–63 days after injection. All attempts to infect salmon fry and rainbow trout in freshwater by oral or intraperitoneal administration of a *Dermocystidium* culture failed. However a subclinical infection did develop in one of six salmon smolts given an intraperitoneal injection of 0.2 ml of a two day-old Oxoid thioglycollate culture. The fish were maintained in sea water and examined 49–129 days after injection.

Discussion

The source of the outbreak of *Dermocystidium* described in this study was presumably wild fish in the river system supplying the farm. Campbell (1974) recorded *Dermocystidium* from brown trout in Scotland and certainly both brown trout and salmon were abundant in the river system at the study site.

The prevalence of the disease on the farm remained low throughout the period of study and there was no evidence of rapid spread of the parasite from fish to fish in individual populations. Clinically diseased salmon were able to survive for at least three months in the laboratory although on the farm such fish eventually died. It may be that all those fish which developed clinical signs became infected from an outside source over a limited period of time.

It is interesting that the disease was only apparent during the winter months when water temperatures were falling or at a low level (<10 °C). This corresponds with the in vitro growth of the parasite, which was greatest at 10 °C. Allen et al. (1968) reported that the virulence of *Dermocystidium* in chinook salmon was greatest at water temperatures below 18 °C. However, the exact role of temperature in influencing the course of the disease is not yet clear.

The stage of the parasite infective to fish was not determined from the transmission experiments. Only vegetative cells were present in the lesion homogenates used and while it is probable that they could initiate a new infection if injected intraperitoneally into a new host, it is not clear if they could directly infect fish if injected into the stomach, or whether they develop into a further invasive stage.

The disappearance of *Dermocystidium* from the farm was surprising and no certain explanation can be offered. In summer 1978 the water intake to the farm was extensively modified so that instead of drawing water from a fairly static reservoir the intake was directly from the fast-flowing river. These changes may have resulted in less favourable conditions for the transmission of the parasite.

The *Dermocystidium* described in this study differs from previously described members of the genus infecting fish by its internal rather than epithelial (fins, skin, gills) infection site (Reichenbach-Klinke and Elkan, 1965; Pauley, 1967). There is evidence that some species, including that discussed here, can become systemic and be dispersed through various organs, but extensive parasite growth is usually restricted to a specific site. Even when the visceral fat tissue of Atlantic salmon was heavily infected macroscopic accumulations of the parasite were only rarely visible in other organs such as the gills.

In view of the massive proliferation of *Dermocystidium* cells normally found in the body cavity of affected fish the minimal tissue reaction by the host was remarkable and in marked contrast to the subacute inflammatory reaction, granulation response, cellular infiltration and hypertrophy described by Pauley (1967) for *Dermocystidium* gill infection in chinook salmon. However the duration of the natural infections histologically examined from the present study was not known and the poor host reaction could be due to very rapid parasite development such as occurred in vitro when nutrient and temperature conditions were suitable.

Dermocystidium from Atlantic salmon is morphologically similar to previously described members of the genus and in common with all others possesses a characteristic cellular stage containing a single large cytoplasmic vacuole (signet-ring) with a prominent vacuoplast (volutin body) (Mackin, 1962; Perkins, 1974). The sequential development of cells leading to the formation of rosette stages through nuclear division, cytoplasmic cleavage and the liberation of daughter cells, which may vacuolate to form the signet-ring stage, closely parallels the developmental cycle of *D. marinum*, the only species for which the life cycle has been described (Perkins, 1969).

Ultrastructure of the composition and formation of the cell walls, appearance of mitochondria and the structure of the nucleus, endosome and vacuoplast was similar to that described for vegetative stages of *D. marinum*. The affinity of vacuoplasts with the outer plasmalemma in sections of some cells has not been previously reported in *Dermocystidium* and in these cases a possible association with lomasomes may be postulated. As with *Dermocystidium* from Pacific salmon (Perkins, 1974) no centrioles have been observed in the material from Atlantic salmon. The failure to obtain development of motile stages made it impossible to determine its exact affinities with other *Dermocystidium* spp. and particularly *D. marinum*.

The economic significance of the disease caused by *Dermocystidium* in Atlantic salmon is not clear. Although a relatively small proportion of fish on the farm were affected, the high unit cost of juvenile salmon means that even low-level mortalities can cause disproportionately heavy economic losses. An additional factor is that salmon smolts from the study farm are shipped to other farms throughout Scotland, thus increasing the chance of spreading the disease. Further studies on the disease should particularly concern the epidemiology of the parasite and methods of control.

References

Allen RL, Meekin TK, Pauley GB, Fujihara MP (1968) Mortality among chinook salmon associated with the fungus *Dermocystidium*. J Fish Res Board Can 25:2467–2475
Campbell AD (1974) The parasites of fish in Loch Leven. Proc R Soc Edinb Ser B 74:347–364
Levine ND (1978) *Perkinsus* gen. n. and other new taxa in the protozoan phylum Apicomplexa. J Parasitol 64:549
Mackin JG (1962) Oyster disease caused by *Dermocystidium marinum* and other micro-organisms in Louisiana. Publ Inst Mar Sci Univ Tex 7:132–299
Mackin JG, Owen H, Collier A (1950) Preliminary note on the occurrence of a new protistan parasite, *Dermocystidium marinum* n. sp. in *Crassostrea virginica* (Gmelin). Science 111:328–329
Pauley GB (1967) Prespawning adult salmon mortality associated with a fungus of the genus *Dermocystidium*. J Fish Res Board Can 24:843–848
Perkins FO (1966) Life history studies of *Dermocystidium marinum*, an oyster pathogen. Ph D Thesis, Florida State University, pp 272
Perkins FO (1969) Ultrastructure of vegetative stages in *Labyrinthomyxa marina* (=*Dermocystidium marinum*), a commercially significant oyster pathogen. J Invertebr Pathol 13:199–222

Perkins FO (1974) Phylogenetic considerations of the problematic thraustochytriaceous – labyrinthulid – *Dermocystidium* complex based on observations of fine structure. Veröff Inst Meeresforsch Bremerhaven Suppl 5:45–63

Perkins FO (1976) Zoospores of the oyster pathogen, *Dermocystidium marinum*, I. Fine structure of the conoid and other sporozoan – like organelles. J Parasitol 62:959–974

Ray SM (1952) A culture technique for the diagnosis of infections with *Dermocystidium marinum* Mackin, Owen and Collier in oysters. Science 116:360–361

Reichenbach-Klinke H, Elkan E (1965) The principal diseases of lower vertebrates. Academic Press, London New York, pp 600

The Causative Agent of Proliferative Kidney Disease May Be a Member of the Haplosporidia

C. SEAGRAVE, D. BUCKE, and D. ALDERMAN[1]

Introduction

Proliferative kidney disease (PKD) is now regarded as an important disease. It has a high incidence in Europe, and is reported to cause large mortalities in farmed rainbow trout fingerlings *(Salmo gairdneri,* Richardson; Ferguson and McAdair, 1977; Ghittino et al., 1977; de Kinkelin and Gerard, 1977; Ferguson and Needham, 1978; Seagrave and Bucke, 1979). A protozoan parasite always observed in the viscera of affected fish is widely regarded as the probable causative agent (Ghittino et al., 1977; Ferguson and Needham, 1978). In this report the organism will be referred to as "PKX" (proliferative kidney disease organism "X").

A similar disease was originally described by Plehn (1924), who suggested the organism was an amoeba and termed the disease "amoebiasis". Recently, Ghittino et al. (1977) and Ferguson and Needham (1978) have also described the organism as an amoeba. However, our own ultrastructural studies of the organism have enabled us to reinterpret previous work, and to recognise many similarities between PKX and an important group of oyster parasites, the *Marteilia* (Sporozoa: Haplosporidia). The genus *Marteilia* includes the species *M. refringens,* which causes "Aber" disease in the oyster *Ostrea edulis* (L). This is currently the most economically important shellfish disease in Europe.

Ultrastructural Comparisons

A detailed ultrastructural examination of PKX was performed, and the results were compared to those from studies made by Ferguson and Needham (1978). Comparisons were also made to species of *Marteilia* (in oysters) described by Grizel et al. (1974), Perkins (1976), and Perkins and Wolf (1976). Two major characteristic features of PKX were selected for detailed comparison:

Internal Cleavage

Marteilia sporulate through a unique process of internal cleavage (endogenous budding) as described by Perkins (1976). Some features, such as the process of initial secondary

---

[1] Ministry of Agriculture, Fisheries and Food, Directorate of Fisheries Research, Fish Diseases Laboratory, Weymouth DT 4 8UB, Dorset, England

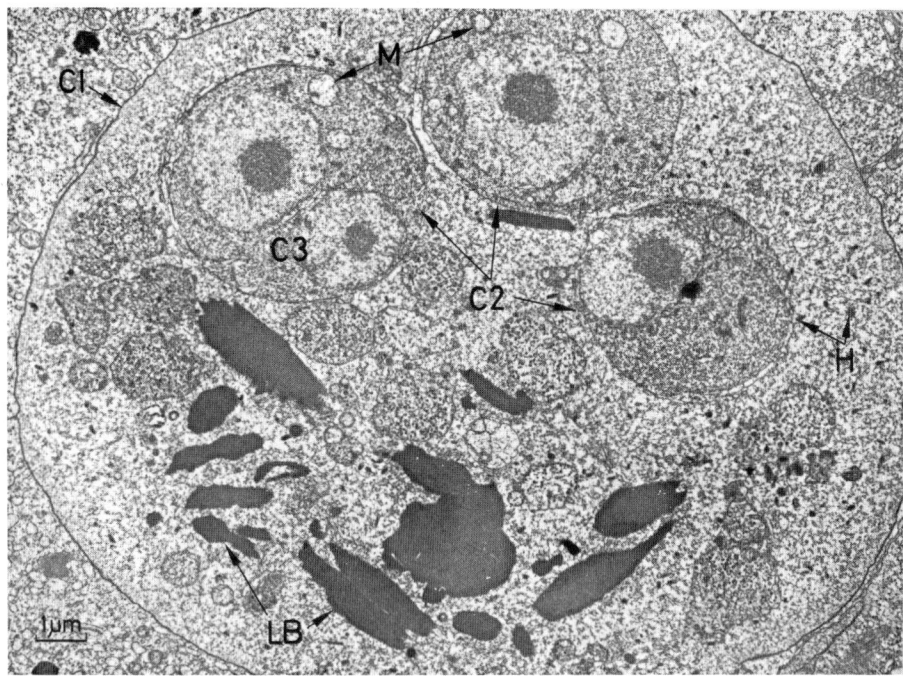

Fig. 1. Immature stage of *Marteilia refringens* with 3 presporangia (secondary cells) and a developing tertiary cell. (Note: this stage of development in *Marteilia* is comparable to stage *(c)* in Fig. 3). *C1* primary cell; *C2* secondary cell; *C3* tertiary cell; *H* haplosporosomes; *M* mitochondria; *LB* laminate bodies (unique in the genus to *M. refringens*). x 9,000 (Photomicrograph courtesy of S. Moss)

cell formation, remain to be elucidated, but the basic process involves delimitation of the "plasmodium" (primary cell) by internal cleavage into uni-nucleate segments which become presporangia (secondary cells). These then enlarge to form sporangia (approximately eight in number). The sporangial protoplast then cleaves and the nucleus divides to produce three or four spore primordia (tertiary cells) which in turn develop into spores. Figure 1 shows an immature plasmodial stage containing presporangia (2° cells).

In their study of PKD, Ferguson and Needham (1978) described "inclusion cells" of unknown significance within the cytoplasm of PKX. Following our own studies we now suggest that these cells are the initial stages of a similar sporulation sequence to that found in *Marteilia*.

Using the terminology of Grizel et al. (1974) for *Marteilia*, we refer to these inclusion cells in PKX as secondary cells, and the stages within these as tertiary cells.

In PKX the secondary cells appear to increase in number as the cell matures. [Ferguson and Needham (1978) have noted up to six.].

Occasionally, tertiary cells may be observed within these secondary cells, but indications of spore development within this tertiary stage have not been recognised.

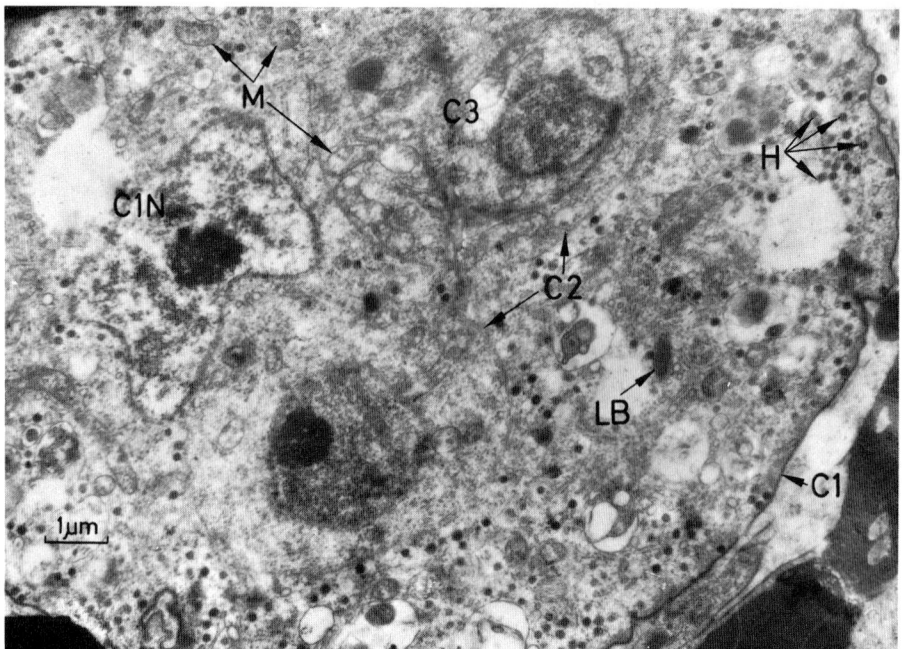

Fig. 2. An "advanced" stage of PKX with two secondary cells and a developing tertiary cell. (Note: this stage of development in PKX is comparable to stage *(c)* in Fig. 3) *C1N* primary cell nucleus. x 10,600

Figure 2 shows a PKX cell with two secondary cells, one of which contains a tertiary cell.

The development processes within *Marteilia* and PKX are compared in Fig. 3. It can be seen that PKX does not attain the same degree of development as *Marteilia*. The sporulation sequence appears very similar up to the stage at which the *Marteilia* secondary cells enlarge, and the tertiary cells develop into spore primordia. With PKX it is generally more common to find just one to three secondary cells, and only occasionally an indication of tertiary cell formation.

"Haplosporosomes"

The *Marteilia* are characterised (as are other members of the Haplosporidia) by the presence of small electron-dense bodies dispersed throughout the primary cell cytoplasm (Fig. 1 and Fig. 4). These are not found, however, in the secondary or tertiary stages, but do "re-appear" in the outermost protoplast of the developing spore. Perkins (1976) describes these bodies as membrane-bound electron-dense inclusions, differentiated into a cortex and medulla, the regions being separated by a membrane which appears as an electron-lucent zone. Their size range is as follows:

    M. refringens   130–370x50–130 nm
    M. sydneyi   146–603x29– 60 nm

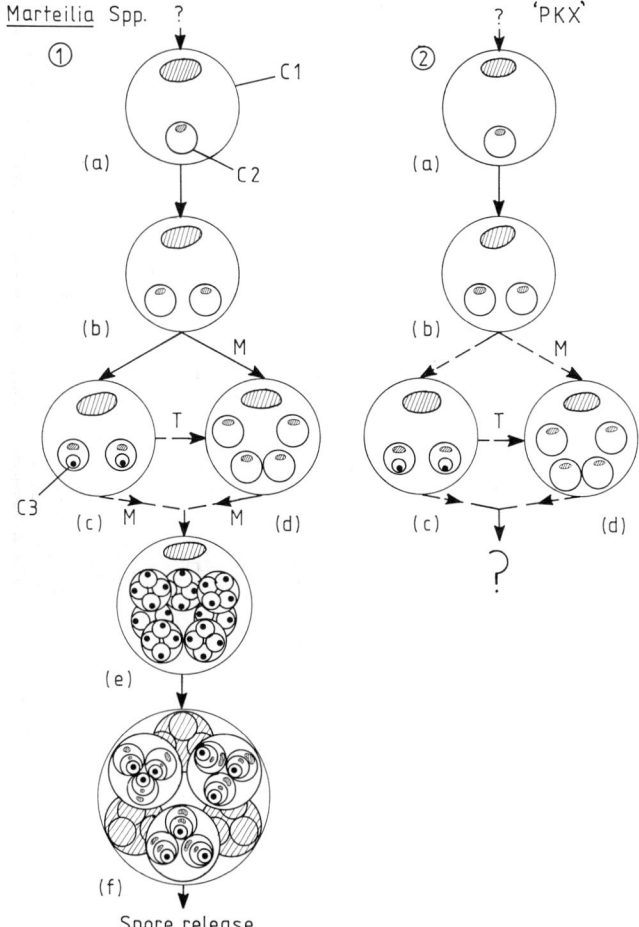

Fig. 3. The possible sporulation sequences found in *Marteilia* spp. and PKX. This sequence may be similar in both up to stage *(e)* (development of spore primordia). Note: dotted lines indicate that the exact "route" in the development process is unclear. *C1* primary cell; *C2* secondary cell; *C3* tertiary cell; *M* stages of secondary cell multiplication; *T* stage of possible tertiary cell migration into primary cell cytoplasm (this could be the mechanism of secondary cell multiplication)

Their shape is usually vermiform or oblate; however, throughout the Haplosporidia in general there is much diversity in this aspect, although the basic structure is almost identical.

Similar bodies are found in PKX (Fig. 2 and Fig. 5) and, as in the *Marteilia*, they are only found in the primary cell cytoplasm, not in the secondary or tertiary stages. These bodies are usually spherical and surrounded by a membrane which appears as an electron-lucent zone. They also contain an electron-lucent "bar" (Ferguson and Needham, 1978). This bar, however, may be an invagination, since in some of the bodies the outer membrane appears to be continuous within the bar. The size range of these bodies in PKX is 140–200 nm.

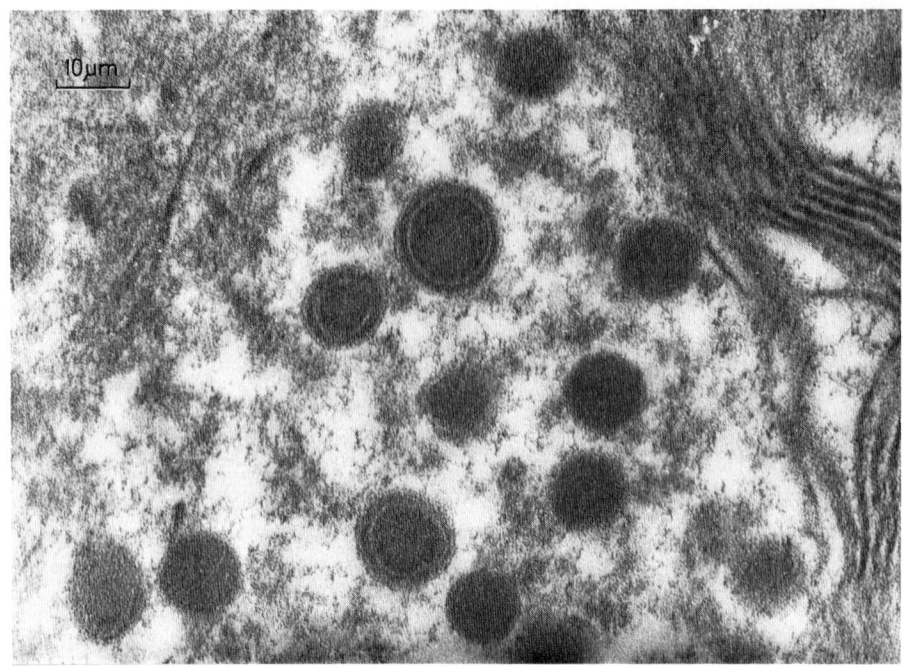

Fig. 4. Haplosporosomes from *M. refringens.* x 130,000 (Photomicrograph courtesy of S. Moss)

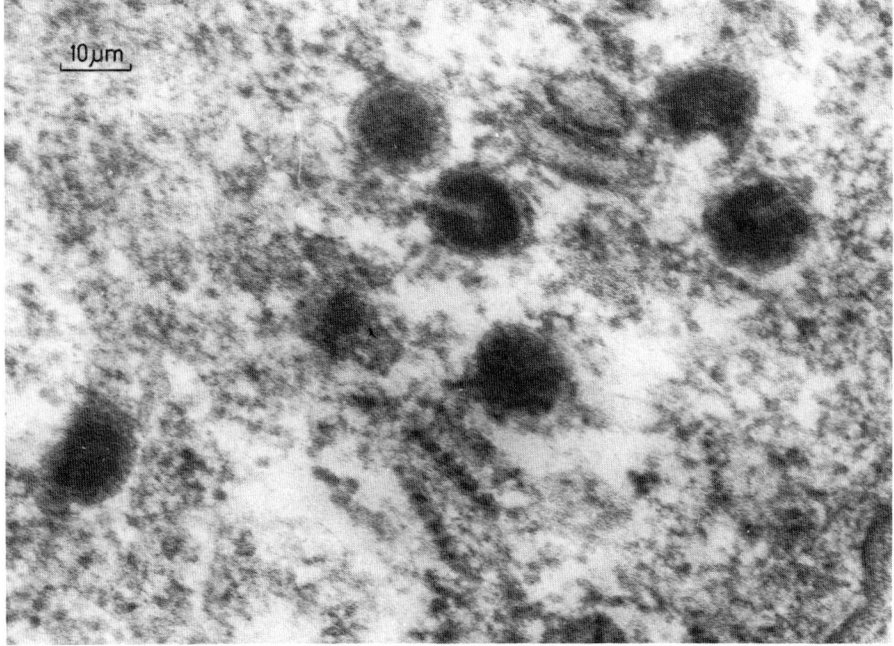

Fig. 5. Haplosporosmes from PKX. x 120,000

Discussion

The proposed relationship between PKX and the *Marteilia* is based upon the two major characteristics of the *Marteilia* group, namely internal cleavage and the presence of haplosporosomes. The internal cleavage of PKX shows many similarities to that found in *Marteilia*, but a *full* sporulation sequence has not been observed. This, however, is not an uncommon phenomenon in the Haplosporidia. For example, *M. refringens* will develop to the secondary cell stage and no further in the *stomach* of the oyster. It will only complete sporulation in the digestive gland. Furthermore, another Haplosporidian parasite, *Minchinia nelsoni* in *Crassostrea virginica* sporulates only rarely, with development normally stopping at the plasmodial stage (Sindermann, 1974). Thus, if PKX is related to this group it would not be unusual to observe it in a non-developing state only.

In PKX an advanced tertiary stage can occasionally be observed, but it can be interpreted in two ways (see Fig. 6). First, that the cell is starting to sporulate and that the tertiary stage is a developing spore primordium or second, that the tertiary cell, having been formed by internal cleavage within the secondary cell, will migrate out and into the primary cell cytoplasm to appear as another secondary stage. The tertiary

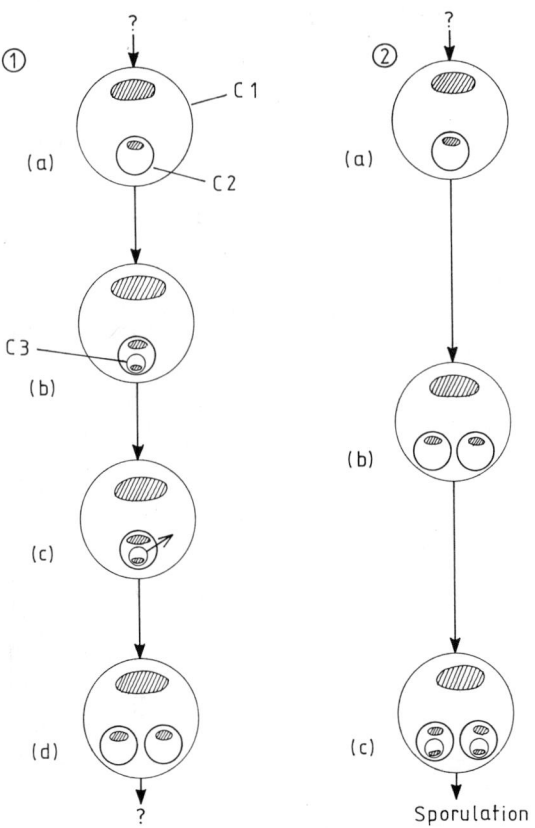

Fig. 6. Two possible development sequences for the tertiary cell stage in PKX. In *1* the tertiary cell migrates into the primary cell cytoplasm to become another secondary cell, whereas in *2* the tertiary cell may be the developing spore primordia

stage could therefore be regarded as the mechanism of secondary cell formation. Thus the exact significance of the tertiary stages in PKX is not clear, and may only be elucidated after further studies.

The electron-dense bodies in the primary cell cytoplasm of PKX show many of the characteristics of the haplosporosomes as defined by Perkins (1976). Each member of the Haplosporidia has its own unique shape of haplosporosome, although their internal structure is almost identical. Originally, these bodies were termed "virus-like particles" (Grizel et al., 1974), but to date their nature and function remain unknown.

Perkins (1976) placed the *Marteilia* in the Haplosporidia on the basis that haplosporosomes plus internal cleavage appeared to be unique to this group. However, Perkins (personal communication) has since drawn our attention to work by Current and Janovy (1977) on a species of Myxosporidia containing bodies with haplosporosome characteristics. He has also noted a similar type of internal cleavage within the group.

However, attitudes as to the inter-relationships between the various organisms currently placed in the Haplosporidia and the Myxosporidia are presently in a state of flux, as more evidence becomes available. We feel at present that PKX certainly shows characteristics of the *Marteilia*, but further work could possibly demonstrate other affinities. We would not wish to make at this stage any proposals with possible taxonomic implications for *Marteilia* or indeed the Haplosporidia as a whole. Thus, on the basis on the similarities described in this text, we believe that PKX may be related to the *Marteilia*. There does not appear to be any good ultrastructural evidence to suggest that there is a relationship to the amoebae (which presumably refers to members of the Sarcodina) as suggested by other workers. Further planned work involves more ultrastructural examination for possible developing spore stages and the screening of other animals (particularly invertebrates from "affected" water supplies) in the search for a possible alternate or true host for the organism, in an attempt to elucidate the complete life cycle of PKX.

References

Current WL, Janovy J Jr (1977) Sporogenesis in *Henneguya exilis* infecting the channel catfish: an ultrastructural study. Protistologica T XIII, Fasc 2:157–167

Ferguson HW, McAdair B (1977) Protozoa associated with proliferative kidney disease in rainbow trout. Vet Rec 100:158–159

Ferguson HW, Needham EA (1978) Proliferative kidney disease in rainbow trout. J Fish Dis 1:91–108

Ghittino P, Andruetto S, Vigliani E (1977) The amoebiasis of hatchery rainbow trout. Riv It Piscic Ittiop XII No. 3:74–89

Grizel H, Comps M, Bonami JR, Cousserans F, Duthoit JL, le Pennec MA (1974) Reserche sur l'agent de la maladie de la glande Digestive de *Ostrea edulis* Linne. Science et Pêche, Bull Inst Pêches Marit No 240:7–30

de Kinkelin P, Gerard JP (1977) Reunion sur l'hepatonephrite parasitaire de la Truite Arc-en-Ciel. Bull Off Int Epiz 87(5–6):489–490

Perkins FO (1976) Ultrastructure of sporulation in the European flat oyster pathogen, *Marteilia refringens* – taxonomic implications. J Protozool 23(1):64–74

Perkins FO, Wolf PH (1976) Fine structure of *Marteilia sydneyi* sp. N Haplosporidian pathogen of Australian oysters. J Parasit 62(4):528–538

Plehn M (1924) Prakticum der Fischkrankheiten. E Schweizerbart, Stuttgart

Seagrave CP, Bucke D (1979) Water threat to fingerlings. Fish Farmer 3:28–29

Sinderman CJ (1974) Diagnosis and control of mariculture diseases in the United States. Middle Atlantic Coastal Fisheries Center, National Marine Fisheries Service, Technical Series Report No 2, pp 306

Poster Section

Preparation of Salmonid White Blood Cells for Virological Studies

P.-J. ENZMANN[1]

We have recently applied the fluorescent antibody technique for rapid detection of viral antigens in several organs of infected trouts as well as in the supernatant of infected cell cultures (Enzmann, 1978, 1979). For diagnosis of nonacute infections of trout with VHS virus a method was developed which allows the rapid detection of viral antigens in leukocytes after isolation of the white blood cells. Leukocytes have been shown also to become infected during an acute VHS infection and may be functionally impaired during the disease. Fluorescent antibody studies on the persistence of infection and on the detection of antigens in acute infections will be described.

Leukocytes from the peripheral blood were isolated by two steps of centrifugation in density gradients. The blood cells were centrifuged first into a cushion of "Uromiro" followed by two density-gradient centrifugations in "Percoll". Four bands could be detected containing from top to bottom: (1) degraded lymphocytes, (2) degraded granulocytes; lymphocytes, (3) granulocytes and lymphocytes, (4) granulocytes.

About 10^3 cells from each band were centrifuged on to microscope slides and then stained with fluorescent antibodies.

Isolated leukocytes from uninfected trouts could be infected and assayed for virus after about 40 h. In band 3 lymphocytes with a specific fluorescent pattern were visible. Isolated leukocytes from trouts infected 6 weeks previously and showing no signs of disease were analyzed for virus directly after isolation. As above, several cells containing viral antigens could be detected in the cells from band 3.

The capacity of viruses to persist in lymphocytes and travel to various tissues in the body as well as the possibility of affecting the immunocompetence of such cells may have significance for the pathogenesis of acute, persistent, and recurrent viral disease.

References

Enzmann P-J (1978) DVG, Fachgruppe Tierseuchenrecht, S 181–183
Enzmann P-J (1979) Acta Virol 23:329–334

---

1 Federal Research Institute for Animal Virus Diseases, P.O.Box 1149, D-7400 Tübingen, FRG

Use of Immunoperoxidase Technique for Detection of Fish Virus Antigens

M. FAISAL and W. AHNE[1]

Introduction

Viral diseases of fish, such as hemorrhagic septicemia of trout (VHS), infectious pancreatic necrosis of trout (IPN), and spring viremia of carp (SVC), are common in European fish cultures. The diagnosis of these diseases is usually based on isolation and identification of virus by neutralisation test.

The fluorescent antibody technique (FAT) has also been used for detection of fish virus antigens (Jørgensen and Meyling, 1972; Piper et al., 1973; Pfeil, 1978). However, autofluorescence of fish tissues can frequently interfere with the interpretation of the diagnostic results. The immunoperoxidase technique (IPT) provides advantages over FAT, e.g., the high specificity, elimination of background staining, the use of an ordinary light microscope, and the keeping of preparations for a long time (Kurstak, 1971).

In fish virology the IPT was used first by Nicholson and Henchal (1978) for the detection of IPN-viral antigen in Atlantic salmon (AS) cells.

The present study was started in order to use the IPT as a diagnostic tool for detection of the following fish-pathogenic viruses in cell cultures and fish tissue (RVC):

    Egtved virus   (causing VHS)
    IPN virus   (causing IPN)
    Rhabdovirus carpio RVC   (causing SVC).

Materials and Methods

Cell Culture

PG cells (Ahne, 1979) were grown in plastic flasks (75 cm^2), petri dishes, leighton tubes, and microtiter plates at 20 °C using MEM supplemented with 10% fetal bovine serum and antibiotics.

Viruses

The viruses used were Egtved virus, F_1 (provided by P.E. Jørgensen, Arhus, Denmark), IPN virus isolated from pike (Ahne, 1978b), and RVC (Lab strain 10/5/77). The viruses

---

[1] Institute for Zoology and Hydrobiology, Ludwig Maximilians University, Kaulbachstraße 37, D-8000 München 22, FRG

were replicated in PG cells at 15 °C (Egtved and IPN) and at 20 °C (RVC). Titration of the viruses were carried out in microtiter plates ($TCID_{50}$) or using the plaque technique (pfu) (Wolf and Quimby, 1973).

Antisera

Antisera against Egtved virus, IPN virus, and RVC were prepared in rabbits as described elsewhere (Ahne, 1978a).

Prepration of Samples for FAT and IPT

Confluent monolayers of PG cells grown in leighton tubes (1x3.5 cm coverslips) were infected with the 3 viruses used in a multiplicity of 1 pfu/cell. Infected cells were removed 2, 4, 6, 8, 10, 12, and 24 h after infection and samples (3 each) were prepared for titration of cell-associated virus, FAT and IPT. For FAT and IPT infected cells were fixed with cold acetone (−20 °C) for 10 min.

CHSE-214 cells infected persistently with IPN virus (Ahne, 1977) were fixed in the same way.

Twenty carp (35 g) infected with RVC by the water route (4×10^4 pfu/ml H_2O) at 13 °C were taken 20 days after infection for isolation of virus from pools of organs, frozen sections (4–6 μ) of gills, kidney, liver, spleen, and heart were prepared for IPT.

Examination of PG Cells and Carp Tissue for Endogenous Peroxidase

Fixed cell cultures grown on coverslips and the frozen sections of carp tissue were inoculated for 5–10 min at room temperature with the peroxidase-specific substrate (Graham and Karnovsky, 1966), which consisted of a saturated solution of 3,3' diamino-benzidine-tetrahydrochloride (Fluka) in 0.05 M tris HCl buffer (pH 7.6) with 0.01% H_2O_2.

Removal of Endogenous Peroxidase

The sections of tissue of carp were immersed in 0.074% hydrochloric acid in ethanol for 15 min at room temperature. Afterwards they were incubated in a humified chamber under a drop of 0.85% sodium phosphate buffer (pH 6.8) for 3 h (Weir et al., 1974).

Indirect Immunoperoxidase Technique

Fixed coverslip cell cultures infected with Egtved virus, IPN virus, and RVC as well as noninfected cells were incubated with specific antisera for 1 h at 37 °C in a humified chamber. Afterwards samples were washed 3 times with 0.05 M tris HCl buffer (pH 7.6), air-dried and incubated with a 1/20 dilution of sheep antirabbit IgG-horseradishperoxidase conjugate (Pasteur Institute, Paris) at 37 °C for 1 h. After washing 3 times with tris buffer the samples were incubated with the substrate as mentioned above. Frozen sections of noninfected and RVC-infected carp were treated in the same way using RVC antiserum.

Immunofluorescence

Indirekt immunofluorescence staining of coverslip cultures was made parallel to IPT according to the methods described by Ahne (1978a).

Results

Normal PG cells showed no endogenous peroxidase activity. Control cells such as noninfected PG cells treated with antisera and virus-infected cells treated with normal rabbit serum exhibited a very faint, unspecific brown coloration of the cytoplasm (Fig. 1a).

PG cells infected with RVC showed a positive IPT reaction 2 h p.i. In contrast first immunofluorescence was evident 6 h p.i. and CPE first appeared more than 24 h after infection.

IPN-virus-infected PG cells gave the first IPT reaction about 4 h p.i., while the fluorescence was positive 6 h p.i. and the CPE was evident 36 h p.i. PG cells infected with Egtved virus showed positive IPT and FAT reaction 8 h p.i. whereas CPE first appeared 24 h after infection.

All infected cells prepared for IPT showed the specific diffuse faint brownish discoloration in the cytoplasm and the perinuclear area exhibited definite globular deep-brown bodies (Fig. 1b).

With the progress of multiplication of the viruses the infected cells exhibited degeneration and the cytoplasm was filled with intense brown granules. The nuclei were always free of the IPT reaction (Fig. 1c, 1a,b).

No crossreaction could be observed using heterologous antisera.

In the case of CHSE cells persistently infected with IPN virus, 50% of the cells gave a positive IPT staining.

Since noninfected tissue of carp treated with the substrate only showed a nonspecific IPT reaction, endogenous peroxidase of all samples used was removed according to the method described by Weir et al. (1974).

After such treatment only samples of kidney and spleen of RVC-infected carp showed a positive IPT reaction (Fig. 2c), whereas the tissue of infected and noninfected carp was negative. The virus titers of samples of kidney and spleen of infected carp were 10^8 $TCID_{50}$/g. The other tissue (liver, heart, gills) was free of positive IPT reaction in this study.

Discussion

The results show that both FAT and IPT are specific methods for detection of fish virus antigens in infected cell cultures and in tissue of virus-infected fish. Using cell

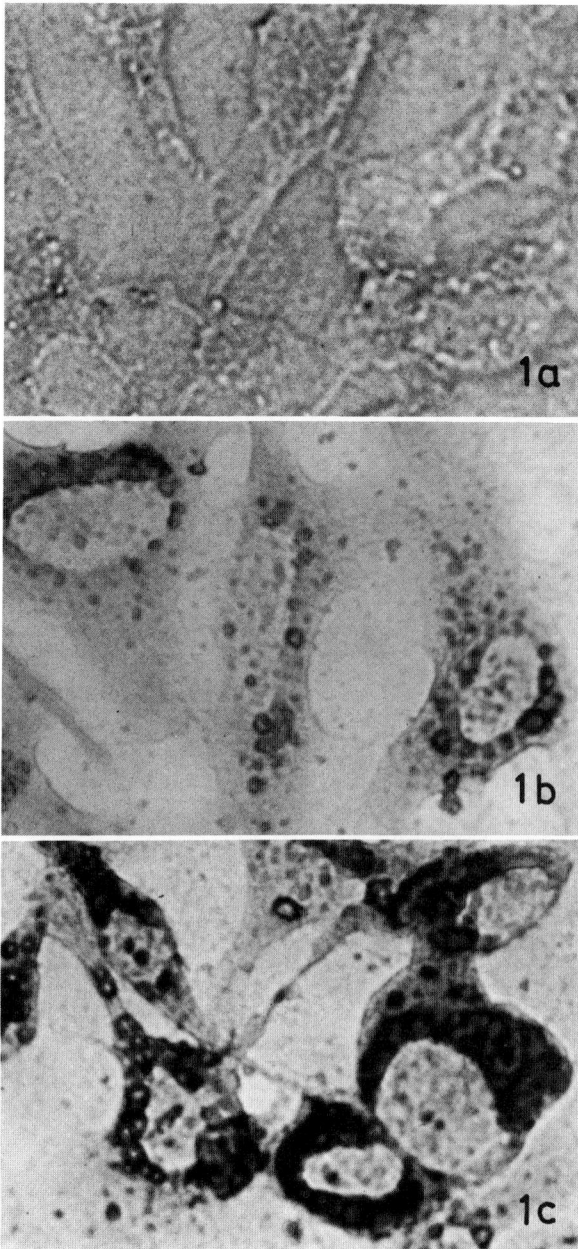

Fig. 1 a–c. Demonstration of RVC antigen in infected PG cells by the indirect immunoperoxidase method. **a** Noninfected PG cells. x 1,000; **b** 12 h p.i. The antigen is localized around the nucleus. x 1,000; **c** 24 h p.i. strong staining of virus-containing cytoplasm. x 1,000

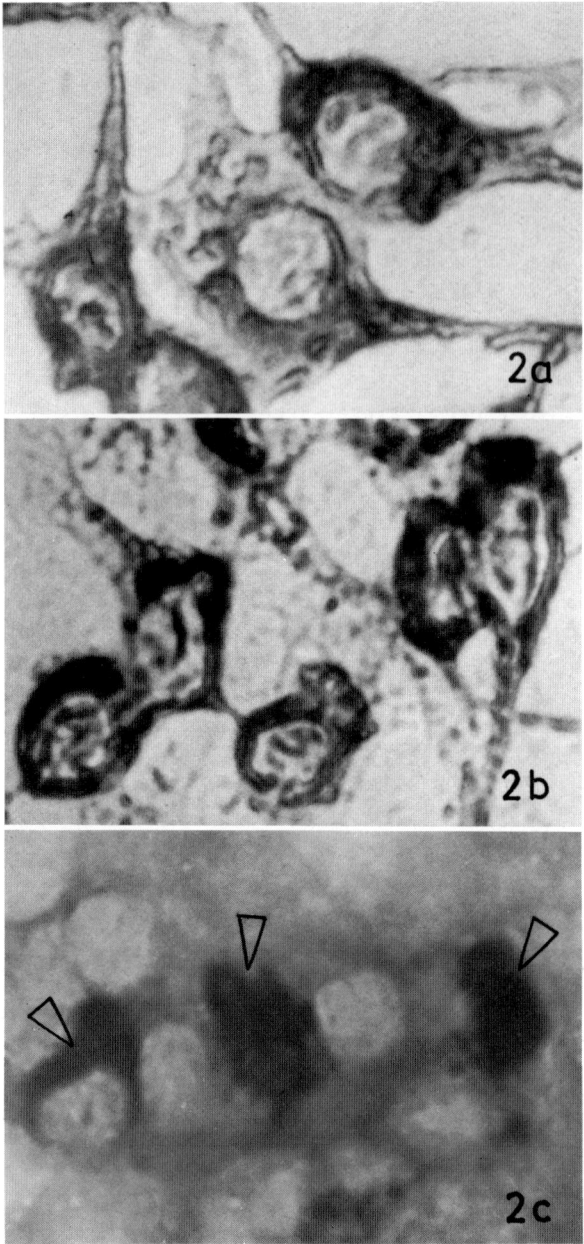

Fig. 2 a–c. Demonstration of VHS and IPN antigens in infected PG cells (**a, b**) and detection of RVC antigen in the kidney of carp experimentally infected with RVC (**c**). **a** VHS antigen in PG cells 24 h p.i. x 1,000; **b** IPN antigen in PG cells 24 h p.i. x 1,000; **c** RVC antigen in the kidney of RVC-infected carp *(arrows)*. x 1,000

cultures the IPT may be considered to be slightly more sensitive because the virus antigen was detectable earlier with IPT than with FAT in the case of RVC and IPN viruses. These findings are in agreement with results obtained with other viruses, described by Kurstak (1971), Bellon et al. (1975), Boorsma et al. (1976), and Faisal (1977).

Our preliminary results demonstrate that IPT is a suitable and rapid method for detection of fish virus antigen in tissue culture. This is in agreement with results of Nicholson and Henchal (1978).

The unspecific positive IPT reaction we have obtained using tissue of carp may be due to endogenous peroxidase present normally in the tissue. Similar reactions have been found in the spleen of rabbit (Weir et al., 1974) and in human nasal secretory cells (Gardner et al., 1978). The method we used to remove such undesired reactions proved to be effective for carp tissue. However, our preliminary results show that viral antigen could be detected in RVC-infected tissue of carp. The stage of infection and the kind of tissue as well as the part of organs used for diagnosis can influence the results. In order to establish the IP technique for detection of virus antigens in tissue of infected fish further investigations will be undertaken.

Acknowledgment. We thank C. Held for her technical assistance.

References

Ahne W (1977) Persistent infection in CHSE-214 cells with IPN virus isolated from pike (Esox lucius). Bull Off Int Epiz 87:415–416
Ahne W (1978a) Laboratoriumsdiagnostik fischpathogener Viren. Tierärztl Umschau 33:584–594
Ahne W (1978b) Isolation and characterization of Infectious Pancreatic Necrosis Virus from pike (Esox lucius). Arch Virol 58:65–69
Ahne W (1979) Fish cell cultures: A fibroblastic line (PG-cells) from ovaries of juvenile pike (Esox lucius). In vitro 11:839–840
Bellon B, Sapin C, Druet P (1975) Comparaison de la sensibilité des techniques d'immunofluorescence et d'immunoperoxydase en methodes directe et indirecte. Ann Immunol 126 C:15–22
Boorsma DM, Streefkerk JG, Korns N (1976) Peroxidase and fluorescein isothiocyanate as antibody markers. A quantitative comparison of two peroxidase conjugates prepared with glutaraldehyde or periodate and a fluorescein conjugate. J Histochem Cytochem 24:1017–1021
Faisal M (1977) Studies on some biological properties of the Quaranfil virus. Master Thesis Fac Vet Med, Univ of Cairo, Egypt
Gardner PS, Grandien M, McQuillin J (1978) Comparison of immunofluorescence and immunoperoxidase methods for viral diagnosis at a distance: a WHO Collaborative study. Bull WHO 56:105–110
Graham RC, Karnovsky MJ (1966) The early stages of absorption of injected horseradish peroxidase in the proximal tubules of mouse kidney: Ultrastructural cytochemistry by a new method. J Histochem Cytochem 4:291–301
Jørgensen PEV, Mayling A (1972) Egtved virus: Demonstration of virus antigen by the fluorescent antibody technique in tissues of rainbow trout affected by viral haemorrhagic septicaemia and in cell cultures infected with Egtved virus. Arch Gesamt Virusforsch 36:115–122
Kurstak E (1971) The immunoperoxidase technique: Localization of viral antigens in cells. In: Maramorosch K, Koprowski H (eds) Methods in virology, vol V. Academic press, London New York, pp 423–444
Nicholson BL, Henchal EA (1978) Rapid identification of infectious pancreatic necrosis virus in infected cell cultures by immunoperoxidase technique. J Wildl Dis 14:465–469

Pfeil C (1978) Modell der Schnelldiagnose einer Infektion mit dem Rhabdovirus carpio durch die indirekte Immunofluoreszenz. Fisch Umwelt 5:17–19

Piper D, Nicholson BL, Dunn J (1973) Immunofluorescent study of the replication of infectious pancreatic necrosis virus in trout and atlantic salmon cell cultures. Infect Immun 8:249–254

Weir EE, Pretlow TG, Pitts A, Williams EE (1974) Destruction of endogenous peroxidase activity in order to locate cellular antigens by peroxidase labelled antibodies. J Histochem Cytochem 22:51–54

Wolf K, Quimby MC (1973) Fish viruses: Buffers and methods for plaquing eight agents under normal atmosphere. Appl Microbiol 25:659:664

Nephrocalcinosis of Rainbow Trout (Salmo gairdneri Richardson) in Freshwater; a Survey of Affected Farms

J.G. HARRISON[1]

Introduction

Nephrocalcinosis in cultured rainbow trout *(Salmo gairdneri)* has been described by a number of authors since Wunder (1967); most recently by Harrison and Richards (1979). Landolt (1975) and others have compared nephrocalcinosis with visceral granuloma of brook trout *(Salvelinus fontinalis)*, and suggested that the two diseases may be species variants.

Experimentally it has been shown that renal calcification can be induced in this species by a low-magnesium diet (Cowey et al., 1977) and also that elevated levels of carbon dioxide in the ambient water produce renal and stomach lesions resembling those seen in spontaneous outbreaks of nephrocalcinosis, the lesions increasing in incidence and severity with increased CO_2 levels (Smart et al., 1979). O'Brien and McArdle (1977) and Ferguson and Needham (1978) report renal calcification associated with proliferative kidney disease (PKD). There have, hitherto, been no data on common factors between affected farms. A survey of farms with suspected nephrocalcinosis was therefore undertaken in the light of these reports.

Methods

The survey was confined to commercial stocks on fish farms and no experimental groups were included.

Diagnosis was by gross and histopathological examination, using methods described by Harrison and Richards (1979). Histopathological material was not collected from all farms. Where other salmonid species were kept on the farms, these were sampled using the same methods. A brief history was taken for each case. The nature of the water supply (e.g. river water, spring, borehole, etc.) was ascertained and any available geological and hydrological data recorded. Use was made of previous water analysis data. A minimal "on-site" water analysis consisted of measurements of free CO_2, total hardness, calcium and magnesium hardness, pH and temperature. Where practicable samples were returned to the laboratory for more detailed analysis. All analytic methods were based on those in Taras et al. (1971). Where practicable repeat samples of water and fish were obtained from time to time.

---

[1] Unit of Aquatic Pathobiology, Stirling University, Stirling, Scotland

A sample of each of six successive groups of fry reared at Farm A (see Table 1) was obtained and examined histologically to determine the incidence of lesions. Farm O, with a known history of PKD associated with occasional mild renal calcification, was included in the survey.

A group of fry from Farm A, transferred to a site known to be free of nephrocalcinosis, were sampled on transfer and at approximately monthly intervals thereafter. Fish of market size at farms K and L, which had been obtained as fry from Farm D, were sampled. Fourteen farms which were known to be free from nephrocalcinosis

Table 1. Water source, résumé of water analysis and diet of surveyed farms. Farms A–J had a problem diagnosed as nephrocalcinosis. Farms K and L received affected fry from Farm D but other groups showed no evidence of nephrocalcinosis. Farm M received fry from Farm A. Farm O had a history of PKD associated with renal calcification

Farm	Water source	Ca^{2+} (mg/l)	CO_2 ppm	Diet	Notes
A	Spring	90	18–20	1	See Table 3 for group variations
B	Spring-fed river	130	15	1	
C	Spring-fed river	96	15	1	
D	Spring	120	13	2	Fry supplied to Farms K and L
E	Borehole	39	20	2	
F	Spring	120	60	3	
G	Borehole	96	45	3	
H	Spring and Artesian	120	55	3	Brown trout on this farm not affected by nephrocalcinosis
I	Borehole	32	22	4	Alkali dosing system later reduced CO_2 to 0
J	Spring-fed river	108	22	3	
K	Spring-fed river	41	6	2	Only fry from Farm A affected
L	River	24	4	2	Only fry from Farm D affected
M	Lake	10	1	5	Only fry from Farm A affected
N	Springs and Borehole	120	4	4	Aeration used to reduce CO_2 from 23–48 ppm level
O	Stream	7	2	4	PKD present

were used as controls. Feeding practices were noted and an analysis fo the food obtained from the manufactur.

Results

Table 1 is a résumé of water analyses and also indicates the diets used. At each of Farms A–J, renal calcification diagnosed as nephrocalcinosis was observed. The severity of lesions increased at higher CO_2 levels. At Farms K, L, and M nephrocalcinosis was seen only in the groups brought in from Farms A and D. At Farm O renal calcification was seen in a few 0+ and 1+ fish but gastric lesions were absent. At Farm N and all the other sites surveyed nephrocalcinosis was absent. At Farm I, where the supply has a natural CO_2 level of 22 ppm, a lime-dosing system was installed which reduced this level to 0 ppm. Groups of fry reared after this showed no signs of renal or gastric calcification. At Farm N, the water originated from a number of boreholes and natural springs with a CO_2 level of 23–48 ppm. The mixed water from these sources was vigorously aerated, resulting in a fall in CO_2 levels to 4–11 ppm. Nephrocalcinosis was absent from two groups of 50 fish (mean weights 3 and 8 g) which were examined both grossly and histologically.

Amongst fry transferred from Farm A to Farm M the incidence of gross nephrocalcinosis fell from 4% to 0% during the growing period, and that of histopathological damage from 15% to 0%. Fry transferred from Farm D to Farm L showed a healing pattern similar to the latter group and gross signs were absent at harvest. At Farm K, however, fry from the same groups still showed gross signs (incidence 10%–20%) at harvest; histologically stomach lesions were absent and renal lesions were confined to the major collecting ducts and ureters. There was histopathological evidence of healing in these groups and this process will be described in a separate communication.

The results of the manufacturer's feed analysis are presented in Table 2. The results of the survey of fry at Farm A are presented in Table 3.

Discussion

The most striking finding of the survey was that all farms where nephrocalcinosis was diagnosed had a CO_2 level in excess of 12 ppm and that the severity of the lesions in-

Table 2. Four different commercial rations were fed on the farms in the survey and the control farms. At Farm M fish are fed a mixture of diet 2 and trash fish in a varying proportion. Dietary constituents are expressed as % of dry matter

DIET	Ca	Mg	P (as PO_1)
1	1.6	0.2	1.2
2	2.0	0.3	1.5
3	2.3	0.5	0.5
4	3.0	0.2	1.3

Table 3. Successive groups of fry from Farm A were sampled at approximately 18 weeks after hatch. Water conditions are very uniform (temp. 9°–10°C. CO_2 18–20 ppm) and feeding is ad libitum for all groups. Fry were partially transected, fixed and then each fish divided into three transverse blocks, thus presenting sections of the kidney at three points and one section through the stomach

Group	Sample Size	Total affected	Pairs of groups compared by arcsin transformation (P, significance level. NS, not significantly different)					
			43	77	81	19	18	20
43	50	4	–					
77	42	24	P>0.001	–				
81	51	15	P>0.01	NS	–			
19	24	13	P>0.001	NS	P>0.05	–		
18	51	0	P>0.01	P>0.001	P>0.001	P>0.001	–	
20	45	4	NS	P>0.001	P>0.01	P>0.001	P>0.01	–

creased with the CO_2 level. Furthermore, the lowering of CO_2 concentrations to below this level, either by aeration or alkali dosing, prevented the appearance of lesions. At all the farms reported, the high CO_2 level was a natural property of the inflow water; however, in superintensive systems employing oxygenation metabolic CO_2 would also contribute to the level in the water and thus might be expected to exacerbate the problem. In the cases reported here the flow rates were such that CO_2 levels at inflow and outflow were not measurably different. The majority of affected farms reported have a rather hard water supply, but that high water calcium per se does not cause nephrocalcinosis is indicated by a comparison of Farms I and N.

The healing of nephrocalcinosis lesions has not been previously confirmed. It seems probable that the rate of healing and its completeness may be influenced by the water calcium level. The occurrence of healing when fish are moved to a low-CO_2 environment is further evidence for the primary importance of CO_2 in the pathogenesis of the disease. The absence of nephrocalcinosis in brown trout at Farm H indicates that this species may resist levels of CO_2 which induce lesions in rainbow trout. A visceral granuloma-like syndrome was observed in brook trout at Farm I which had not been exposed to high CO_2 levels. Parallel groups of rainbow trout had no lesions. Similar cases have been observed at two other farms with CO_2 levels of less than 1 ppm. It therefore appears that this syndrome (which will be described more fully elsewhere) is different in aetiology from nephrocalcinosis.

The diets used on all the farms except Farm M were standard commercial rations which were also used on the control farms. The levels of magnesium in all rations are greatly in excess of these used by Cowey et al. (1977) and it seems unlikely that diet plays a significant role in the pathogenesis of the lesion.

The wide variations in the incidence of lesions amongst fry reared in essentially similar conditions (Table 3) may indicate a variation in susceptibility at least to moderate CO_2 levels. Clearly if this were heritable it might offer hope of a long-term solution to the problem. Unfortunately, the group with the lowest incidence also grew the most slowly. Obviously several factors besides inheritance could explain these data.

The low CO_2 level at Farm O indicates that renal calcification following PKD is not related to nephrocalcinosis.

Acknowledgements. This work was funded by the Natural Environment Research Council and Shearwater Fish Farming Ltd. My thanks are due to Drs. R.H. Richards and G.R. Smart for their help and advice, to the managements of all the farms involved and to Dr. J. Lasserre and Dr. P. de Kinkelin for arranging visits to a number of farms in France.

References

Cowey CB, Knox D, Adron JW, George S, Pirie B (1977) The production of renal calcinosis by magnesium deficiency in rainbow trout *(Salmo gairdneri).* Br J Nutr 38:127–135

Ferguson HW, Needham EA (1978) Proliferative kidney disease in rainbow trout, *(Salmo gairdneri,* Richardson) in fresh water. J. Fish Dis 1:91–108

Harrison JG, Richards RH (1979) The pathology and histopathology of nephrocalcinosis in rainbow trout *(Salmo gairdneri,* Richardson) in fresh water. J Fish Dis 1:91–108

Landolt M (1975) Visceral granuloma and nephrocalcinosis of trout. In– Ribelin , Migaki (eds.) Pathology of fishes. University of Wisconsin Press, pp 1004

O'Brien DJ, McArdle J (1977) A renal disease of possible protozoan aetiology resembling nephrocalcinosis in rainbow trout. Ir Vet J 31(3):46–48

Smart GR, Knox D, Harrison JG, Ralph JA, Richards RH, Cowey CB (1979) Nephrocalcinosis in *Salmo gairdneri* Richardson: the effects of prolonged exposure to elevated free carbon dioxide concentration. J Fish Dis 2:279–289

Taras MJ, Greenberg AE, Hoak RD, Rand MC (eds) (1971) Standard methods for the examination of water and waste water, 12th ed. American Public Health Assoc, Washington, pp 1193

Wunder W (1967) Peculiar renal changes and the formation of a renal cyst in the rainbow trout. Allg Fisch Z 95(15):470–471 (Translated, Multilingual Service Division, Ministry of Environment, Canada, No 676711)

The Increase of Nephrocalcinosis (NC) in Rainbow Trout in Intensive Aquaculture

H.-J. SCHLOTFELDT[1]

Spontaneous nephrocalcinosis (NC, urolithiassis) has been described in several species of salmonids in the wild (Herman, 1971; Landolt, 1975: Harrison, 1977) as well as in sea-farmed and freshwater-farmed rainbow trout (Mulcahy et al., 1980; Schlotfeldt, 1980). This can be confirmed from our own experiences. Different degrees of nephrocalcinosis were more or less regular additional findings in traditionally and extensively kept trout as well as being found sporadically in other salmonids from the wild, but in negligible percentages with no economic importance.

The number of trout intensive aquaculture facilities in Lower Saxony has been considerably enlarged during the last three years. In this Federal Province the development of different flowing trough and recycling technologies was especially enforced. The regular health and sanitary control of some of these units was assumed by the Fish Diseases Control and Health Service of Hannover. In the course of time mentioned above an unexpected increase of nephrocalcinosis has been observed.

The more intensive the facility, the higher was the percentage of NC-diseased fish. This was especially the case when three main factors were coincident:
1. Intensive holding of fish stocks up to 25–30 and more kg body weight/m^3 H$_2$O
2. Artificial oxygen-supersaturation supply up to 50 mg O$_2$/l
3. Borehole (well) water supply (Harrison, 1977; Harrison, personal comunication; Schlotfeldt, 1978).

From such an aquaculture plant monthly samples of trout were investigated during nine months, with the main aim to establish within which period of time apparently healthy fish of the 12–15-cm class would develop first signs of NC. Because of a complex interaction of several disadvantageous technical factors which could not be improved soon, this plant showed extremely unfavorable environmental conditions.

The conditions can be summarized as follows: the flow rate of water through the fish troughs was only 1.7 l/kg body weight/h on average with fish stocks from 25 kg/m^3 upward. Return or recycling of water was not technically possible. With such a minimal flow rate heavily loaded water had to be accepted with values as follows: ammonia up to 11 mg/l; nitrite up to 1.0 mg/l; nitrate up to 8.5 mg/l; Fe 1.0 mg/l; pH 6.8–7.3; temperature 8°–13°C; averages of 12–14 mg O$_2$/l (original oxygen-supersaturation of $\mp$ 40 mg/l).

Additionally an extremely high bacterial contamination of the water was recorded with amounts of up to 6x10^5 of different bacteria such as *Aeromonas hydrophila* and

---

[1] Fish Disease Control and Health Service of Hannover FSGD (Fisch-Seuchenbekämpfungs- und Gesundheitsdienst Hannover) in cooperation with the Fish Disease Research Unit, School of Veterinary Medicine, Bünteweg 17, D-3000 Hannover 71, FRG

A. liquefaciens, Erwinia, Mesentericus subtilis group, *Flavobacteria, Colibacteria, Staphilococcus albus* and other Staphilococcae as well as several other ubiquitous, non-pathogenic, gram-negative and gram-labile *Coccus*-like bacteria which could not be differentiated biochemically.

With monthly clinical and histopathological sampling an attempt was carried out to follow the development of the NC syndrome in a freshly arrived trout population of the 12–15-cm class (weight average ∓27 g) from the beginning until the consumption size (∓28 cm/300 g). The investigated population started with 60,000 trout. Batches of 17–20 fish were examined monthly. Sampling commenced 2 days after arriving at the plant. The whole number of fish was of the same origin. Their hygienic and sanitary conditions were known and were — as far as it is possible to state — unobjectionable. Water supply and oxygen-supersaturation technology were the same for all troughs. With differential virological, bacteriological, and other diagnostic methods nephropathic syndromes like viral hemorrhagic septicaemia (VHS), infectious pancreatic necrosis (IPN), bacterial kidney disease (BKD), mycotic infections (ichthyophoniasis, *Phoma herbarum*), and tumors (metastasis of hepatomes, nephroblastomes) were excluded. As monthly controls 5 trout were taken from another intensive aquaculture unit which showed the same oxygen-supersaturation supply but satisfactory water-flow rates of 9–10 l/kg/h and similarly appropriate water values.

The clinical development can be summarized as follows: the described environmental conditions necessitated reduced feeding, the only way in which heavier losses could be avoided. Feeding was orientated according to short-term slight improvements of water quality. Reduced feeding resulted in unsatisfactory weight increase. Stress such as transport was poorly tolerated with losses above average. After 3 months — with only smal losses — the following symptoms were observed, which increased slowly: one part of the fish population showed retarded growth, others did not reduce their growth while showing clear symptoms. Fish demonstrated swelling of the abdomen, exophthalmos of varying degrees without periorbital hemorrhages (Fig. 1) occasional skin petechiae, hemorrhages at the fin base, fin damage caused by the high density of the population, protruded anus. Inside a high degree of ascites was observed: watery and anemic muscular tissue; different consistencies and color changes of the liver, which was yellowish, marmorea and sometimes hemorrhagic; splenomegalia; often pericarditis; dilation and inflammation of the stomach, gastroenteritis. Depending on the stage of the disease changes of the excretory (caudal) part of the kidney became evident. Initially the kidney showed dilation and sinuosity of ureters with only slight enlargement of the organ (Fig. 2). Advanced cases of the syndrome showed different stages of calcinosis from calcium grit up to visible "kidney stones" with 2–3 times enlargement of the organ (Fig. 3). These extreme changes were evident after the seventh month in the plant.

From the 8th to 9th month on ward up to 20% of the fish were affected by the symptoms described. Because of poor-looking, slow, and stunted growth and low value of flesh due to edema, these fish carcasses were considered inedible and not suitable for processing purposes like smoking. The plant was forced to destroy them and therefore the disease became economically important.

In contrast to NC cases observed in Canada (Fig. 4; Kelly and Yurkowski, Freshwater Institute Winnipeg, Manitoba, personal communication, 1978) our observations

Fig. 1. Severe exophthalmos without periocular hemorrhages of NC-diseased trout of the 15–18-cm class

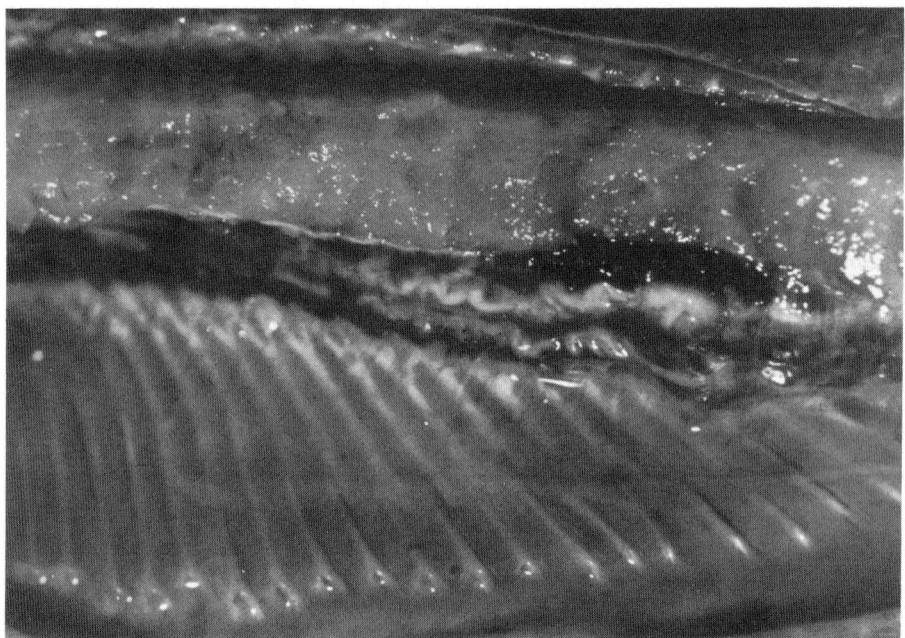

Fig. 2. Beginning NC showing dilation and sinuosity of ureters with only slight enlargement of the kidney

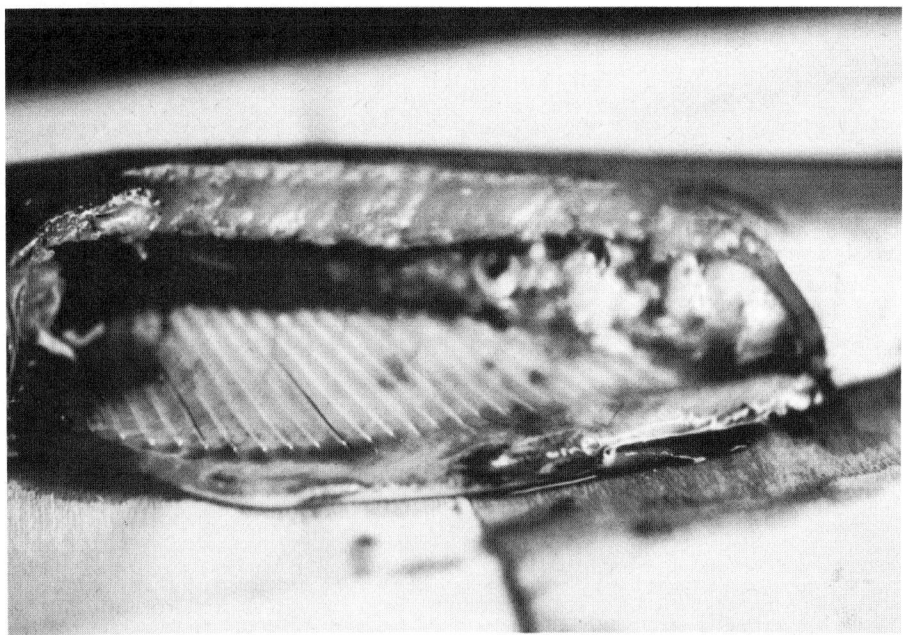

Fig. 3. Advanced stage of NC with complete calcinosis and 2–3 times enlargement of the posterior (caudal, excretory part) kidney

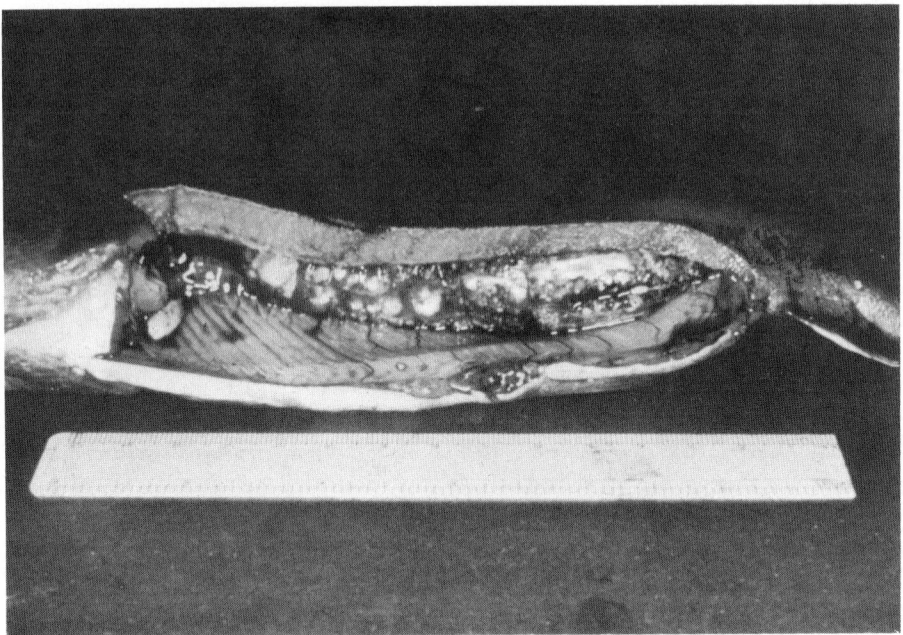

Fig. 4. NC observed in Canada with calcinosis of the anterior and posterior kidney (by courtesy of R.K. Kelly and M. Yourkowski, Freshwater Institute, Winnipeg, Manitoba, 1978)

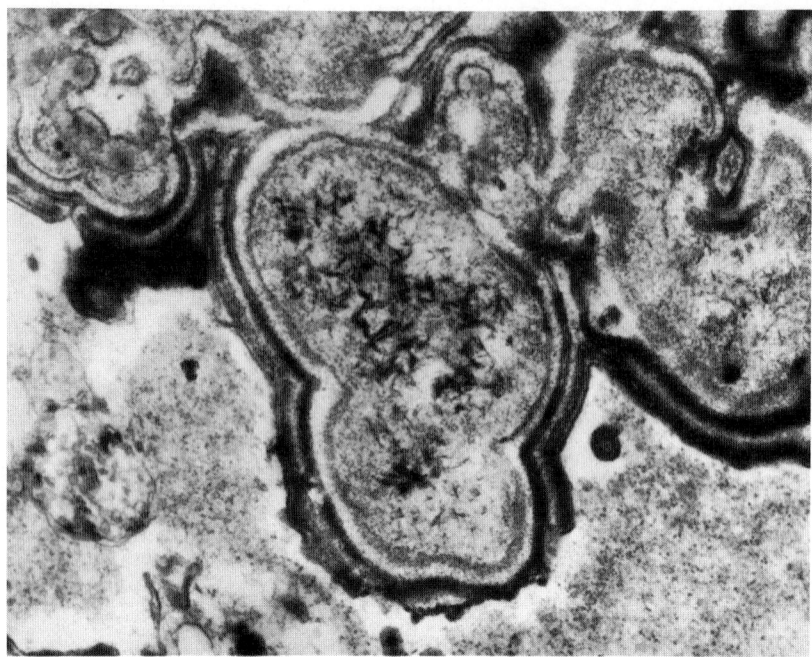

Fig. 5. Electron micrograph of thin section of brook-trout kidney showing the lamellar structure of calcium oxalate deposits within a tubule. The small needle-like crystals within to oxalate are probably calcium phosphate (apatite). By courtesy of T. Yamamoto, Edmonton, Canada)

in question showed the kidney deterioration confined strictly to the caudal (excretory) part of the organ.

In spite of massive kidney damage and severe alterations of the liver — in the absence of additional stress factors — mortality remained low. Calcium concrements were chemically identified as mainly calcium phosphate. Recent electron-microscopic and histochemical studies of NC granules in brook- and rainbow trout by Yamamoto (personal communication, 1979) have shown that different morphological materials are present. The concentric lamellar structures often found in kidney stones of higher animals are present. These concentric forms are known to consist largely of calcium oxalate and appear to be initiated by many foci and increase in size to coalesce and form larger stones. Another material observed in electron-microscopic studies is composed of many needle-like crystals that aggregate together. These deposits are thought to be composed of calcium phosphate (apatite) and have been observed to be present inside calcium oxalate crystals or adjacent to them (Fig. 5; by courtesy of Prof Dr. T. Yamamoto, Edmonton, Canada). The monthly histopathological examination showed on the one hand alterations without macroscopic kidney changes, and premature proliferative gastritis on the other hand, which appeared considerably earlier than kidney alterations did. Kidney samples of the first two months were macroscopically and microscopically negative.

During the third month up to 30% of fish showed incipient NC with only slight anatomic and pathological changes. In the course of the fourth month NC alterations were observable on 70% of fish. The fifth (monthly) sample showed slight decrease; from the sixth sample on, however, an increase of changes was again observed.

At first the histopathological lesions were intratubular cylindrical calcareous bodies in the distal tubular segments of the kidney. Later, interstitial reactions and granulom-like nephritis also developed in the proximal segments with extensive intratubular calcareous concretions and intertubular metastatic and dystrophic sclerosis. The stomach showed initially few alterations, but later developed inflamed sclerotic foci. This was followed by granuloma-like gastritis of the mucosa and submocosa. Furthermore, and especially in heavy changed stomachs with damage of the mucosa and submucosa, multiple, round and/or oval "onion-like" lamellar bodies with sporadic central sclerotic foci were observed.

These structures showed great similarity to "mycosis-like granuloma" as described by Wood et al. (1955) and Wood and Yasutake (1956). These gastric alterations were regularly observed in NC-diseased fish, but also in isolated single cases without any kind of nephritic change. These findings show the systemic character of the calcium-metabolic alteration with primary extrarenal manifestation and possibly the cause. The occasional liver examination showed histological and clinical symptoms of fatty degeneration. A relationship between this finding and the lesions of stomach and kidney was not evident in every case analyzed. In our cases it was not possible to point out sclerotic foci in the muscular tissue as described by Harrison and Richards (1979).

Recent studies of Harrison and Richards (1979) and Smart et al. (1979) pointed out the importance of increased CO_2-concentrations as a possible or additional NC cause. These authors were able to induce NC-similar conditions experimentally with high CO_2 concentrations in the water (Harrison, personal communication, 1978). This factor was not measured in the present case and could have had some influence. On the other hand NC was until now not evident in other intensive aquaculture facilities controlled by our Health Service over more than 3 years and which show the same conditions of water and oxygen supply as well as high fish stocks, but enough water to ensure a minimal flow rate of more than 7 l/kg/h and thus appropriate water quality values.

Unsuitable feeding with lack of magnesium (Cowey et al., 1977) or an excess of cottonseed meal in the fish diet (Dunbar and Herman, 1971) could play a role in the development of NC. But it is improbable that in the present case diet deficiencies could have had an influence, in view of the fact that exactly the same commercially available fish feeds have been fed for years in other intensive aquaculture units (with satisfactory environmental conditions) without having induced any NC symptoms higher than the normal rate. Practical experiences of our Health Service in Lower Saxony over several years showed that the minimum flow rate of water through intensive aquaculture (silo and other fish-holding facilities) units seems to be 5 l/kg body weight/h. But with this amount the slight additional stress induces first depressive symptoms like lack of appetite and indolence. The absolutely lowest limit seems to be 3 l/kg/h, which should be avoided (Müller, personal communication, 1978, 1979). Further observations of our Health Service in several intensive aquaculture plants showed that flow rate values from 7 l/kg/h upward represent an optimum. Extremely dense fish stocks of

75–150 kg body weight/m^3 with short-term loads even up to 185 kg/m^3 (in that case without feeding) but with water-flow rates of 7–8 l/kg/h and more (and hence a tolerable catabolite load) did not develop any substantial increase of NC.

At another silo aquaculture unit with sufficient water flow (10 l/kg/h) and unobjectionable water quality values it was possible to observe that because of short-term unavoidable circumstances silo-raised consumption-sized trout were transferred to a raceway with sufficient but organically high-loaded brooklet water. Fish from the silo were clinically healthy and regularly controlled. Their NC findings, compared to those of fish from natural ponds, were only insubstantially higher. Fish from the loaded raceway showed after 2–3 months raceway-holding a notable increase in frequency as well as in severity of observed NC.

The results of our material showed the decrease of fish health and increase of the NC syndrome under the influence of the extremely unfavorable water quality described and the water-flow rate of only 1.7–1.8 l/kg/h.

In addition to the insufficient water supply and enrichment of nitrogenous catabolites above average, the stress produced by the extreme bacterial contamination of the water has to be considered.

Bacteriological examination of the samples showed bacterial encroachment on internal organs of the fish in question. It is well known that an excess of ubiquitous, initially more or less nonpathogenic bacteria from the water can lead to serious disease conditions with final septicemic symptoms.

Thus, the sum of stress factors described could have led to an increase of the alterations mentioned in this initially healthy and unobjectionable experimental fish group after a period of 6–7 months in the facility. During our observations the NC syndrome showed a clear tendency to develop earlier than after 6–7 months. Furthermore these examinations showed that NC in intensive aquaculture is unlikely to be due to a single factor. It is more likely to be due to a multiplicity of factors in which the complex of environmental conditions is of crucial importance (Mulcahy et al., 1980; Schlotfeldt, 1980).

Without additional stress and in spite of massive damage of the excretory part of the kidney as well as severe alterations of the liver, the mortality was low. This may support the hypothesis that in fish the gills, gut, and parts of the skin are able to take over a major part of kidney functions (Schlotfeldt, 1980).

The survival of trout under the conditions described shows the enormous adaptability of the "domestic" pond-rainbow trout, but does not signify at all that such precarious conditions should be used. During planning of intensive aquaculture facilities the experiences described should be considered. If lack of adequate amount and quality of water could lead to such a sum of extremely disadvantageous conditions the construction of the plant should not be pursued.

Acknowledgments. I gratefully acknowledge the help provided by H.A. Schoon, Institute of Pathology of the School of Veterinary Medicine, Hannover, who carried out and assembled the histopathological observations; to G. Kirpal and G. Amtsberg, Institute of Microbiology of the School of Veterinary Medicine, Hannover, for their bacteriological assistence; to B. Liess, Institute of Virology of the School of Veterinary Medicine, Hannover, in which the virological examinations were formerly carried out by our Service; to H. Rüssel, Institute of Chemistry of the School of Veterinary Medicine, Hannover, who carried out the chemical analysis of the calcium concrements.

References

Cowey CB et al. (1977) The production of renal calcinosis by magnesium deficiency in rainbow trout (Salmo gairdneri). Br J Nutr 38:127–135

Dunbar CE, Herman RL (1971) Visceral granuloma in brooktrout (Salvelinus fontinalis) J Nutr 101:1445–1451

Harrison JG (1977) Salmonid renal histopathology. Proc 2nd Symp FAO/EIFAC. Coop Progr Res Aquacult Fish Dise (COPRAQ), Brest, France, 25.–28.10.1977

Harrison JG, Richards (1979) The pathology and histopathology of nephrocalcinosis in rainbow trout (Salmo gairdneri Richardson) in fresh water. J Fish Dis 2:1–12

Herman RL (1971) Visceral granuloma and nephrocalcinosis. US Bur Sport Fish Wildl Fish Dis Leafl 32:39

Landolt ML (1975) Visceral granuloma and nephrocalcinosis of trout. In: Ribelin WE, Migaki G (eds) The pathology of fishes. The University of Wisconsin Press, Madison Wisconsin, pp 793–801

Mulcahy MF, Collins N, McAuliffe T (1980) Nephrocalcinosis in seafarmed and freshwater farmed rainbow trout in Ireland. ICES Special Meeting on Diseases of Commercially Important Marine Fish and Shellfish, Copenhagen, 1–3 October 1980. Communication No. 21, pp 1–7

Schlotfeldt HJ (1978) Praxisnahe Ergebnisse der 2. Tagung der FAO/EIFAC-COPRAQ (Gemeinschaftliche Forschungsprogramme in Aquakultur und Fischkrankheiten) in Brest, Frankreich vom 25.–28. Okt. 1977. Fisch Teichwirt 2:30–33

Schlotfeldt HJ (1980) Some clinical findings of a several years survey of intensive aquaculture systems in Northern Germany, with special emphasis to gill pathology and nephrocalcinosis. European Inland Fisheries Advisory Commission. XI Session. Symposium on New Developments in the Utilization of Heated Effluents and of Recirculation Systems for Intensive Aquaculture. Stavanger, Norway, 28–30 May 1980. EIFAC/80/Symp.:E/75, pp 1–19

Smart GR et al. (1979) Nephrocalcinosis in rainbow trout Salmo gairdneri Richardson: The effects of exposure to elevated CO_2 concentrations. J Fish Dis 2:279–289

Wood EM, Yasutake WT (1956) Histopathology of kidney disease in fish. Am J Pathol 32:845–853

Wood EM, Yasutake WT, Lehmann WL (1955) A mycosis-like granuloma of fish. J Infect Dis 97:262–267

Trial Vaccination of Rainbow Trout Against Aeromonas liquefaciens

ACUIGRUP[1]

Introduction

Bacterial hemorrhagic septicemia (BHS) produced by *Aeromonas liquefaciens* has been diagnosed in over 20 fish farms for 1973–1976, it being the most common disease of the rainbow trout in Spain (Acuigrup, 1979).

Takin into account the several factors affecting this condition in a negative manner, some experiments have been carried out in 1977–1978 and 1979 to determine whether vaccination produces an effective protection against this disease, and the role played by environmental and nutritional factors. This pathological process has already been described by several authors (Wood, 1968; Wedemeyer, 1970; Wedemeyer and Wood, 1974).

We have preferably chosen mass vaccination by immersion as the protection system for trout against BHS because of immunization results obtained in viral (IHN and SHV) and bacterial (vibriosis, redmouth, etc.) fields.

During all outbreaks registered in 1974–1977 a continuous antibiotherapy was necessary to reduce mortality. Taking into account the economic importance of the disease we tried to look for a good protection system by two routes: immunization and environment control.

Material and Methods

Facilities

Experiments were carried out in Navafria (Segovia) fishfarm. Water coming from the Cega river has temperatures ranging from 2 °–3 °C in winter to 25 °C in summer, its volume being enough to maintain fish in ponds and raceways. In summer, because of scarce volume the water was recycled and aeration was provided by a floating pump running at the rate of 200 m^3/h. In this way the oxygen content could be increased by 1 ml/liter under the most unsatisfactory conditions.

Environment Factors

According to our more detailed study (See Acuigrup, 1979) the factors which exert a negative influence and have been controlled by us are as follows: temperature, oxygen,

---

[1] Laboratorio de Ictiopatologia, Acuicultura Bioter, Apartado, 29008 Madrid 27, Spain

and ammonium. Throughout the year average temperature ranges from 3 °C to 25 °C, and was as follows during the experimental period (May–October): May 15 °C, June 16 °C, July 18 °C, August 23 °C, September 20 °C, October 16 °C for 1976–1978; however, for 1977 the average temperature in August was 17 °C, and 20 °C in September, it being an abnormal summer.

There is limited oxygen in August and September because of high temperature, recyclages, etc., its values ranging from 6.5 ppm on coming into the pond to 4.5 ppm at the exit. The ammonium value is between 0.1 and 0.4 ppm under the worst conditions.

Vaccines

From the microorganism isolated in this fish farm for 1975–1976, whose characteristics (shown in Table 1) are similar to those of *Aeromonas*, germs were maintained in TSB broth in stock until reactivation to get a bacterial sediment by precipitation. Then the germ was killed and fixed at 7% transparency with green filter in a 440-nm colorimeter.

Immersion immunization was carried out in two steps: first, into hyperosmotic solution at 2.5 kg salt/50 liters of water; then into bacterin solution at the rate of 125 ml/50 liters for 1.5 min; this solution was used on five occasions with additional oxygenation and sometimes it was replaced by a new one. Revaccination was carried out in 1977.

Table 1. Morphological and biochemical characteristics of *A. liquefaciens*

Morphology	Gram-negative bacillus –
Pigmentation	–
Motility	+
B galactosidase	+
Arginine DH	+
Lysine DC	+
Ornithine DC	–
Citrate (Simmons)	–
Sulfydryl production	+
Urease	–
Tryptophan DA	–
Indole	–
VP	+
Nitrites	+
Glucose	+
Mannitol	+
Inositol	–
Sorbitol	+
Ramnose	+
Sucrose	+
Melibiose	–
Amigdaline	+
Arabinose	+

Table 2. Chemical composition and vitamins of diets

	Diet 1	Diet 2
Protein	40.5%	41%
Fat	7.2%	7.5%
Ash	11.8%	12.3%
Fibre	4.1%	3.8%
Moisture	9.1%	9%
N.F.E.	21.3%	20.4%
Vitamin A (i.u./kg diet)	2,000	6,000
Vitamin D_3 (i.u./kg diet)	400	700
Vitamin E mg/kg	18	26
Vitamin B_1 mg/kg	12	21
Vitamin B_6 mg/kg	3.2	7.5
Ca. pantothenate	13	36
Vitamin C mg/kg	30	66
Inositol mg/kg	100	260

We used the original bacterin in oil solution with Freund's adjuvant by the intraperitoneal route at 21 ml/kg body weight. Spray vaccination was carried out using compressed powdered bacterin over trouts out of water.

Nutrition

Table 2 shows the characteristics of basic diets.

Fish Batches

In 1977, experiments were carried out with 50–80-g rainbow trouts. These were divided into 3 batches:

Batch 1: 400 fish, immersion vaccination and revaccination one month later.
Batch 2: 800 fish, immersion.
Batch 3: Control group; remaining fish in the fish farm.

Prevaccination of the first batch was conducted by immersing fish into a hyperosmotic solution at the rate of 1.5 kg (ClNa/50 l water for 2.5 min. Then immersion in a container with the bacterin solution under study was carried out. Exposure times to solutions were only changed in batch 2: 1 min with hyperosmotic solution and 1.5 min with vaccine solution. Additional oxygen was supplied by aeration.

During the summer of 1978 experiments were made by using fish of different origin and size, and vaccination routes were also different.

Fish were divided into 4 batches, as follows:

Batch 1: 25,000 fry (3.5 cm, 0.6 g) from our own fish farm, hatched in April. The maximum load was 3 kg/m^2 in raceways. Immersion vaccination and revaccination 25 days later.

Batch 2: 12,000 fry (10 cm, 12 g) from the USA, hatched in January. Concrete longitudinal ponds; load of 5 kg/m^2. Immersion vaccination.
Batch 3: 113 trouts (17 cm, 60 g) from our own fish farm, hatched in May 1977. Concrete ponds at 10 kg/m^2. Intraperitoneal vaccination.
Batch 4, 5, 6: Controls of batches 1–3, in similar ponds at the fish farm, having the same size, origin, and load. Trouts were maintained in a cage inside the pond.

Batches 1 and 2 were vaccinated by immersion into a hyperosmotic solution for 2 min in 40-l water containers, followed by immersion into vaccine solution for 1.5 min.
Batches in 1979 were:

Batch 1: 120 trouts (16 cm). Intraperitoneal vaccine under the same conditions as those in 1978.
Batch 2: 9,000 fry (10 g). Immersion vaccination into bacterin solution for 2.5 min and 1.5 min.
Batch 3: 9,000 fry (10 g). Immersion vaccination into bacterin solution for 2.5 min and 1.5 min. Revaccination 25 days later.
Batch 4: 120 trouts (16 cm) 2.5 min and 1.5 min immersion.
Batch 5: 120 trouts (16 cm). Spray vaccination.
Batches 6, 7, 8, 9, 10: Controls of batches 1–5.

Results

Experiment 1977

During 1977 vaccinated batches were heavier than unvaccinated ones:

	Weight at vaccination time	Weight in October
Vaccinated groups	50–80 g	225 g
Controls	50–80 g	200 g

From the clinical point of view vaccinated batches showed excellent health without disease symptoms. In contrast, unvaccinated batches presented acute symptoms of the disease, high morbidity, and moderate mortality (at the time of maximum temperature). It was necessary to apply drugs in feed according to the antibiogram.

Experiment 1978

During summer 1978 results were as follows:

Batch 1: Growth, morbidity, and mortality were similar to those in control batch and characteristics as to be expected in the development of these sizes on the fish farm.
Batch 2: No mortality was seen; growth 10% above that of control was shown. Control had to be treated with drugs in feed because of an outbreak with clear symptoms of the disease in the middle of August at maximum temperature.

Batch 3: No mortality, best growth and no external and internal symptoms. These fish were reared 30 kg/m^2 rate at maximum temperature from 15 September to 15 October. Then they were exposed to live bacteria solution *(Aeromonas)*. The fish showed fair growth and no external lesions. Control fish were treated with antibiotics in feed because of a mortality increase; lower growth than vaccinated fish was sustained.

Experiment 1979

Batch 1: 7% mortality was recorded as a result of damage produced by the intraperitoneal needle. Development and survival were normal without incidence of infection as was the case in controls, where these was no mortality.
Batches 2, 3: No significant difference from control batch, without pathological incidence or mortality.
Batches 4, 5: No significant differences between vaccinated and unvaccinated fish.
No infectious bacterial process in experimental batches or control group or remaining fish on fish farm was recorded. After six years of treatment, in 1979 medicated food was not necessary.

Discussion

Taking into account the results of 1977–1978, the conclusion in support of a mass vaccination of over 15-cm trouts could be drawn, it being a protection system against BHS during summer and effective for at least 3 months. Lower growth in control batches could be explained by the pathological incidence which reduces their development as well as the subsequent antibiotic treatment.

The efficiency of intraperitoneal vaccination also shows the fish capacity to produce a strong immune response to a single vaccination. It does not seem to be the most suitable method for mass vaccinations; it is convenient for comparative studies.

The apparently contradictory results of 1979 as regards the previous ones could be explained by considering the several environment factors affecting the disease. During 1978–1979, when water oxygenation conditions became better, a general health improvement on the fish farm was achieved. On the other hand, a BHS outbreak was recorded in 1977 where antibiotic treatment was necessary; after this outbreak had been overcome, two trout batches were made, one was given the previous feed (diet 1) and the other batch received diet 2. Three months later a new outbreak ensued but only in the first batch. Because of this situation, taking into account the outstanding role of nutrition, diet 1 was substituted for diet 2 on our fish farm in June 1978.

Under these conditions the following alternatives were available to us:
Either improving the environment and nutrition conditions of trout, or vaccinating fish as a protection against BHS.

According to the experiment results of 1979, it is concluded that it is more efficient to improve environment and nutritional conditions on the trout farm, since no pathological incidence then occurred. Undoubtedly it is very important for a suitable level and balance of vitamins to be present to fulfil trout requirements, since the deficiency syndrome of some vitamins, shown in Table 1, not only reduces the natural fish defences but also produces skin lesions, these being the open door to aqueous germs such as *Aeromonas,* which leads to a generalized severe septicemia.

Besides, results in 1978 contrary to those of 1979 would justify the fact of diet 2 not producing protection when it was given in June 1979, the month in which experiments started; therefore there was no time for the new diet to act, whereas in 1979 feeding trout with diet 2 during all their life gave good protection.

In conclusion, vaccination could be a protection method against BHS where it is not possible to control the factors contributing to the disease, but immunotherapy is not necessary when factors can be controlled. Finally, the initial purpose of these experiments has been achieved, because a greater knowledge of the nature of the disease and the role played by the environment and nutrition factors has been obtained.

References

Acuigrup (1979) La septicemie hemorragique bacterienne chez la — truite en Espagne. Interaction avez l'ecosysteme. Bull Cent Etud Rech Sci Biarritz 12(3):493–500

Buchanan RE, Gibbons NE (1974) Bergey's manual of determinative bacteriology. Williams and Wilkins Co, Baltimore, pp 345

Meuschmann-Brunner G (1978) Aeromonas of the "Hydrophila-Punctata-Group" in fresh water fish. Arch Hydrobiol 83(1):99–125

Meyer FP (1970) Seasonal fluctuations in the incidence of disease on fish farms. Symp Dis Fish Shellfish. Spec Publ Am Fish Soc 5:21–29

Wedemeyer G (1970) The role of stress in the disease resistance of fishes. Symp Dis Fish Shellfish. Spec Publ Am Fish Soc 5:30–35

Wedemeyer G, Wood JW (1974) Stress as a predisposing factor in fish disease. US Fish Wildl Serv FLD 38:1–3

Wood JW (1968) Diseases of Pacific salmon: their prevention and treatment. Washington State Printing Off, Dep Fisheries Hatch Div, pp 29

Flavobacteriosis in Coho Salmon (Oncorhynchus kisutch)

ACUIGRUP[1]

Introduction

Salmon production in Spain has been increasing during the past years. Geographical areas for these cultures are located in estuaries rich in phytoplankton and zooplankton in North Spain. Most of the pathological problems are related to diseases already found in other countries. Among them vibriosis and kidney disease stand out (ACUIGRUP, 1977). Special location conditions under which these cultures are established, as well as the latitude allow certain diseases to occur in a special environment. These diseases interact with the classical ones and little attention has been given to them. One of these diseases is flavobacteriosis: it is produced by a germ whose classification is not yet well known (Bergey's Manual, 1974; Robert, 1978) and it is scarcely mentioned in the world literature.

Its clinical picture has been described only cursorily in freshwater (Brisou et al., 1964) and sea-water species (Meyer et al., 1959). The latter occurred in a "red tide" on the Californian coast, with a high mortality rate. Typical granulomatous lesions either occurred in freshwater or sea-water species; they had slow development. Disease seems to occur when phytoplankton proliferation is increased in the area, it being used by bacteria (Meyer et al., 1959).

Flavobacteria spp. (occurring naturally in sea-water under certain conditions) have been considered difficult organisms for taxonomic classification because of their changeable morphology, nutritive requirements, physiology, and biochemistry. The French school classifies the mobile organisms as *Flavobacteria* and the immobile ones as *Empedobacter,* while American authors classify both types as *Flavobacteria.* Among them two species are pathogenic for fish; *Flavobacteria balustinum* and *F. piscicida.*

In this paper we try to describe the etiology, clinical picture, and treatment of a flavobacteriosis that occurred in Spain during the summer of 1978.

Material and Methods

Practical control of this disease was carried out in Galicia (North Spain) on an industrial fish farm developing three species: *Salmo coho (O. kisutch), Salmo salar,* and rainbow trout (*S. gairdneri* Richardson).

---

[1] Laboratorio de Ictiopatologia Acuicultura Bioter, Apartado, 29008, Madrid 27, Spain

Salmon weights ranged from 500–1,000 g. Maximum temperatures in summer were between 18 °C and 23 °C. Cages showed evident fouling with zooplankton, preferably mussel larvae *(Methylus edulis)* and different types of algae:

Fucus vesiculosus (brown)
Ulva lactuca (green)
Porfira umbilicalis (red)
Enteromorfa comprese (green)
Polisifonia (red)
Hipoglosum woodwardi (red)

Salmon crowding in cages was about 20 kg/m^3.

Samples were taken from anterior kidney and liver and they were sowed in 1.5% CLNa TSA.

Biochemical behavior was tested in the API system, the reading being controlled at 24 and 46 h. Müller-Hinton plates were used for antibiograms.

Germs isolated in 1978 were kept in agar stock at 4 °C for 12 months. During summer 1979, a 30-salmon batch *(Salmo coho)* was inoculated at the dorsal venous sinus. Weights ranged from 800–1,200 g.

Inoculum was prepared from *Flavobacteria* kept in agar stock and reactivated in 1.5% TSB. A suspension of live bacteria was made with saline solution at the rate of 20 ml in 700 mg of wet sediment. Each fish was given 0.5 ml of this suspension. Experimental challenge was carried out in round ponds (83.3 m^3, 5.4 m ∅ and 1 m depth) in August. Initial water temperature was 18 °C varying by ±1 °C over the day.

In this month sows of the algae present in the area were also carried out to prove the possible incidence of germs.

Result

Salmo coho (O. kisutch) mortality was about 20%–25% in summer 1978. *Flavobacteria* could be detected in all fish studied. Bacteria sown into 0.5%, 1.5%, and 3% salt TSA produced an orange pigment, characteristic after 24 h culture, it being increased in 1.5% TSA (Fig. 1).

Tinctorial and biochemic characteristics are shown in Table 1.

The clinical picture is characterized by a generalized septicemia throughout the body; external and internal lesions are quite similar to those of vibriosis; hemorrhages in sides, fin base, eyes, and cephalic region. Internally gastroenteritis, liver congestion, and splenomegaly.

Germ behavior with different drugs is shown in Table 2.

Results obtained with the 30-salmon batch inoculated at the dorsal venous sinus during summer 1979 show a clinical picture characterized by fast disease progression and 100% mortality. Between 24 and 48 h the following lesions appeared:

Externally, uni- and bilateral hemorrhages in eyes; in some cases eyes affected were lost. This sign was the first one, appearing more than 24 h after inoculation.

Cutaneous hemorrhages in sides and base of ventral and pectoral fins.

Internally, hemorrhagic gastroenteritis with petechia in gonads.

No *Flavobacteria* could be detected in sows carried out with algae taking from cages during summer 1979.

Fig. 1. Yellow-orange pigment produced by *Flavobacteria* spp. (1.5% CLNa TSA)

Discussion

Flavobacteriosis has not been considered a disease capable of producing large losses of fish populations on fish farms. Massive mortalities have only been recorded in marine wild fish, this mortality being related to a big quantity of algae used as a support by *Flavobacteria* (Meyer et al., 1959). According to results obtained in the present study evidence is given of high incidence of morbidity and mortality by *Flavobacteria* in salmon culture when several factors occur: high temperature, a lot of fouling (algae and mussel) and high fish concentration per unit volume. We think that among these factors algae determine *Flavobacteria* occurrence, as different authors have shows (Bein, 1954; Berland et al., 1969). During the year 1978 we determined severe fouling in cages, shortage of renewing O_2 and excessive loading; all these things enhanced the sudden outbreak of disease. We thought that these factors really conditioned the *Flavobacteria* appearance as pathogenic germs. This possibility was confirmed in the experiments carried out during the summer of 1979, when under the same conditions (cages and environment) but without fouling (application of antifouling agent to nets) neither mortality nor increase of *Flavobacteria* were observed in salmons and in the few algae present.

We should explain here that fish affected by the disease during the past summer had mortality rates as high as 80%; the 20% surviving population was affected by the above-mentioned flavobacteriosis. During the same year 1978, acclimatiz one-year-old

Table 1. Tinctorial and biochemical characteristics

Characteristics		Reading at 24 h	Reading at 48 h	Observations
Tinction				Gram-negative
Morphology				Bacillary forms with *Diplobacillus colomes*
Pigment		+		Yellow-orange. Non-diffusible
Motility		+		
Oxidase		+		
Gelatin		−		
Glucose	aerobe	−		Without gas
	anaerobe	−		
Lactose	aerobe	−		Without gas
	anaerobe	−		
B Galactosidase		−		
Arginine DH		−	+	
Lysine CD		−		
Ornithine		−		
Citrate (Simmons)		−		
SH2 formation		−		
Urease		−		
Tryptophan DA		−		
Indole		−		
V.P.		+		
Nitrites		−		
Gelatine		+		
Mannitol		+		
Inositol		−	+	
Sorbitol		+−	+	
Ramnose		−	+	
Sucrose		+−	+	
Melibiose		−	+	
Amigdaline		+−	+	
Arabinose		−	+	

Table 2. Germ behavior with different drugs

Drug	Behavior
Chloramphenicol	S
Tetracycline	S
Furazolidone	S
Sulfamerazine	S
Erythromycin	S
Furanace	R
Novobiocin	R

S = sensible
R = resistance

fish and those next to the batches which had suffered from flavobacteriosis were not affected by this disease but they had severe vibriosis. This phenomenon may be explained by an autodefence process of the flavobacteriosis population against *Vibrio anguillarum* while they were susceptible to other pathogenic bacteria. It was the opposite with the one-year-old salmon population, victims of vibriosis but not of flavobacteriosis.

In any case our experiments confirm the existence of a disease related to algae and even associated with several physiological stages of the same algae.

The germ isolated from the outbreak during summer 1978 shows some biochemical and physiological characteristics different from those of the same germ kept in agar stock and reactivated during summer 1979.

Changes have been specially related to pigment production (less pigment activity), motility (immobile), morphology (forming 4—5 joined-element chains).

These conditions have already been observed by several authors (Macleod et al., 1954). These circumstances and other biochemical and physiological characteristics are responsible for the different behavior and uncertain classification of these bacteria.

Contrarly to review literature reports, lesions of this disease have not been characterized by their granulomatous aspect, slow development, and step-wise mortality, but show a generalized hemorrhagic septicemia picture and fast evolution.

This clinical picture appeared in the natural outbreak of 1978 and in that experimentally produced in 1979. This could be due to the fact that under certain environmental conditions (low load of *Flavobacteria*) the patholigical activity was then slower than later, when a high density of bacteria per culture unit is present, and also the fact that *Flavobacteria* strains could be capable of producing a different clinical picture.

Treatment with oxytetracycline reduced mortality greatly, giving evidence of the viability of a therapeutic control.

In conclusion we can say that:
1. Althoug *Flavobacteria* live saprophytically in marine surroundings they can act pathogenically with high mortality when high algae density and unsatisfactory culture conditions coincide.
2. Isolated *Flavobacteria* have great pleomorphism and changeable biochemical and physiological behavior, which makes their diagnosis difficult.
3. More cases and more detailed studies are necessary to investigate the exact pathogenic relationship "germ-algae", according to the alga type, and more convenient physiologicaldevelopment for bacterial growth and pathogenic activity.

References

Acuigrup (1977) New disease in salmo culture in Spain. Bull Off Int Epiz 87(5—6):515—516
Bein SJ (1954) A study of certain chromugenic bacteria isolated from "red tide" water with a description of a new species. Bull Mar Sci Gulf Caribb 4:110—119
Bergey's Manual (1974) Of determinative bacteriology. Williams and Wilkins Company, Baltimore, pp 357

Berland BR, Maestrini SY (1969) Action de quelques antibiotiques sur le developpement de cing espèces de Diatomées en culture. J Exp Mar Biol Ecol 3(1):62–75

Berland BR, Blanchi MG, Maestrini SY (1969) Etude des bactéries associees aux algues marines eu culture. I. Determination preliminaire des espèces. Mar Biol 2:330–333

Brisou J, Tysset C, Vacher B (1964) Recherches sur les pseudomonadaceae. Etudes de deux souches de flavobacterium isolées des poissons d'eau douce. Ann Inst Pasteur: 633–638

Kluge JP (1965) A granulomatous disease of fish produced by flavobacteria. Pathol Vet 2:545–555

Macleod RA, Onofrey E, Norris ME (1954) Nutrition and metabolism of marine bacteria. I. Survey of nutritional requeriment. J Bacteriol 68:680–686

Meyer SP, Baslow MH, Bein SJ, Marks CE (1959) Studies of flavobacterium piscida Bein. I. Growth, Toxicity and ecological consideration. Bact 78:225–230

Robert (1978) Fish pathology. Baillire Tindall, London, pp 318

Hemagglutination Properties of Aeromonas

T.J. TRUST, I.D. COURTICE, and H.M. ATKINSON[1]

Introduction

Aeromonas salmonicida and *Aeromonas hydrophila* are important pathogens of fish. However, although both species cause significant fish losses throughout the world, little is known concerning their virulence mechanisms. Studies with pathogens of other animals have demonstrated that the first and most important step in many infectious processes is the attachment of the pathogen to the epithelial tissues of the host [2]. Indeed some virulent strains have specialized mechanisms of attachment which enhance their ability to cause disease. However, scant attention has been focused on the ability of fish pathogens to adhere to fish tissue cells.

A simple technique that has been widely used to reveal the ability of bacteria to attach to eukaryotic cells is that of hemagglutination [1]. The ability of various sugars to inhibit hemagglutination has also been used to provide information on the nature and specificity of the receptors on the eukaryotic cell surface [4]. In this study we have used hemagglutination to study the attachment specificities of various strains of *A. salmonicida* and *A. hydrophila*.

Materials and Methods

Strains 6, 10, 46, and 69 were strains of *A. hydrophila* provided by H.M. Atkinson (Australia) and strains 413 and 416 were from J.F. Lee (UK). *A. hydrophila* strains 423 and 424 were from B. Dunsmore (Canada), and strains 434 and 435 were from S. Sanyal (India). *A. salmonicida* 201 was from T. Kimura (Japan), 394 was from I. Smith (UK), and *A. salmonicida* strains 436, 437, 438, and 439 were from T.P.T. Evelyn (Canada).

For coagglutination tests, *A. hydrophila* cultures were grown on blood agar at 25 °C for 24 h, while *A. salmonicida* cultures were grown on trypticase soy agar at 25 °C for 48 h. This incubation time minized interference by the self-agglutinating activity of some strains of *A. salmonicida*. Suspensions of bacteria were prepared in Dulbecco phosphate-buffered saline, pH 7.4 (PBS) to yield approximately 1.5×10^9 bacteria/ml. Human group 0 blood was collected immediately before use, as was rainbow trout blood. Blood from other species was obtained from Flow Laboratories.

---

[1] Department of Biochemistry and Microbiology, University of Victoria, Victoria, B.C., V8W 2Y2, Canada

Blood cells were washed three times in PBS and a 3% (v/v) suspension prepared in this saline. In some experiments a 3% (v/v) suspension of baker's yeast was used instead of the blood cells. Slide coagglutinations were performed at room temperature with 20 μl of blood or yeast cells and 20 μl of bacterial cells. A PBS blood or yeast cell control was always included. Slides were gently rocked by hand and strains were considered negative if coagglutination had not occurred within 10 min. The minimum coagglutinating dose (MCD) was measured as the smallest number of bacterial cells/ml giving visible coagglutination in 10 min and the hemagglutinating or yeast coagglutinating power of a culture was calculated as 10^{11} MCD [4]. Sugar inhibition of coagglutination was tested by performing coagglutination tests in the presence of 20 μl of 1% (w/v) sugar. Any sugars inhibiting coagglutination were tested in a series of twofold decreasing concentrations until and endpoint was reached. The minimal coagglutination inhibitory concentration was the lowest concentration completely inhibiting coagglutination by 10 MCD of bacteria.

Results and Discussion

The various strains of *A. hydrophila* were able to hemagglutinate blood from a wide variety of species including rainbow trout (Table 1). The strongest hemagglutinators of trout cells were strains 69, 412, and 413 with hemagglutinating powers of 2×10^4. *A. hydrophila* strains 46, 423, 434, and 435 gave hemagglutinating powers of 1×10^4, *A. hydrophila* 10 gave 1×10^3, and strain 424 was the poorest hemagglutinator with a hemagglutinating power of 7×10^2.

Sugar inhibition studies (to be published) with *A. hydrophila* and human blood revealed a variety of hemagglutinins. Strains 6, 10, 416, and 424 have a hemagglutinin inhibited by L-fucose of D-mannose, strain 434 has a hemagglutinin inhibited by L-fucose and strain 69 has a hemagglutinin inhibited by D-mannose. The hemagglutinin of strains 412, 413, 423, and 435 is inhibited by D-galactose. *A. hydrophila* 46 has two hemagglutinins, one is inhibited by D-galactose and the other is inhibited by D-mannose. In the case of *A. hydrophila* 6, the hemagglutinin is pilus-mediated, but pili were not seen on other strains. In all cases, the activity of the hemagglutinins was destroyed by heating at 55 °C for 5 min. Hemagglutinins were formed during growth at 20 °C, 25 °C, 30 °C, and 37 °C.

Strains of *A. salmonicida* were not as broad in their hemagglutination spectrum as *A. hydrophila*. Although chicken, guinea-pig, rat, and human blood cells were hemagglutinated, horse and rabbit cells were poorly hemagglutinated (Table 1). Bovine and sheep cells were not hemagglutinated by *A. salmonicida* and only strain 436 gave slow hemagglutination with trout blood. The ability of *A. salmonicida* to hemagglutinate fish blood and human blood was quantitated and the results are shown in Table 3. The best hemagglutinator of human blood was 438, but this strain did not hemagglutinate trout blood. The hemagglutination of trout blood by 436 was one-third as strong as its hemagglutination of human blood. Sugar inhibition studies revealed that the various strains of *A. salmonicida* possess a hemagglutinin inhibited by D-mannose or by L-fucose. In the case of strain 436 this was true for trout blood and for human blood. The two sugars were equally efficient in their ability to inhibit hemagglutination.

Table 1. Ability of *Aeromonas hydrophila* and *Aeromonas salmonicida* to hemagglutinate blood from various species

Blood	Strain of *A. hydrophila*												Strain of *A. salmonicida*					
	6	10	46	69	412	413	416	423	424	434	435		201	394	436	437	438	439
Rainbow trout	–	+	+++	+++	+++	+++	++	+++	±	+++	+++		–	–	++	–	–	–
Horse	+++	–	+++	+++	+++	+++	–	+++	+++	++	+++		+	+	+	–	–	–
Bovine	–	–	++	–	++	++	–	+++	–	+	+++		–	–	–	–	–	–
Chicken	+	–	+++	+++	+++	+++	–	+++	++	–	+++		+++	+++	+++	+++	++	++
Guinea-pig	++	+++	+++	+++	+++	+++	–	+++	++	+++	+++		++	+++	++	++	++	++
Rat	–	+++	+++	+++	+++	++	+	+	++	–	+++		++	–	–	–	–	–
Sheep	–	–	+	+++	++	+++	–	+++	–	++	+++		–	++	–	+	+	+
Rabbit	–	–	+++	–	+++	+++	–	+++	–	++	+++		–	+++	–	–	+++	–
Human	+++	+++	+++	+++	+++	+++	+++	+++	++	+++	+++		++	+++	++	+++	+++	+++

– no hemagglutination
+++ hemagglutination of all blood cells

Electron microscopy failed to reveal the presence of pili on any of the hemagglutinating cultures of *A. salmonicida*. Since the hemagglutinating activity was destroyed by heating at 50 °C for 5 min, the hemagglutinin of *A. salmonicida* would appear to be a surface protein. This protein is active over the pH range 4.5–8.5, but does not appear to be formed during growth on blood agar.

Since yeast cells have surface mannans, *A. salmonicida* was also able to coagglutinate yeast. Table 2 shows that strain 438 was also the best coagglutinator of yeast, and that the yeast coagglutination of all strains was inhibited by D-mannose or L-fucose. Indeed these results suggest that the surface protein responsible for yeast coagglutination and for hemagglutination is truly an adhesin. Clearly this common adhesion of *A. salmonicida* recognizes D-mannose or L-fucose on the surface of eukaryotic cells. The poor ability to hemagglutinate trout cells suggests that this *A. salmonicida* adhesin has difficulty gaining access to the appropriate receptor sugar residues on the surface of the trout blood cells.

The ability of the *A. salmonicida* adhesin to recognize either L-fucose or D-mannose residues on eukaryotic cells was confirmed by quantitating the ability of various

Table 2. Hemagglutinating power and yeast coagglutinating power of *Aeromonas salmonicida* and minimal sugar concentrations inhibiting hemagglutination

Strain		Hemagglutinating and yeast coagglutinating power	Minimal sugar concentration inhibiting hemagglutination or yeast coagglutination[a] (mM)	
			D-mannose	L-fucose
201	Trout	–[b]		
	Human	5×10^3	0.2	0.2
	Yeast	1×10^4	0.4	0.4
394	Trout	–[b]		
	Human	1×10^4	0.2	0.2
	Yeast	5×10^3	0.3	0.6
436	Trout	3×10^3	0.2	0.1
	Human	1×10^4	0.2	0.2
	Yeast	5×10^3	0.3	0.6
437	Trout	–[b]		
	Human	1×10^4	0.2	0.2
	Yeast	5×10^3	0.8	0.8
438	Trout	–[b]		
	Human	2×10^4	0.2	0.2
	Yeast	2×10^4	0.3	0.4
439	Trout	–[b]		
	Human	1×10^4	0.2	0.2
	Yeast	5×10^3	0.4	0.4

[a] Determined with bacterial suspensions 10 times the minimal coagglutinating dose
[b] No hemagglutination

Table 3. Sugar inhibition of hemagglutination and yeast coagglutination by *Aeromonas salmonicida* strains 394 and 436

	Minimal sugar concentration inhibiting hemagglutination or yeast coagglutination[a] (mM)			
Sugar	Human blood		Yeast	
	394	436	394	436
L-fucose	0.2	0.2	0.6	0.6
D-fucose	N[b]	N	N	N
p-Nitrophenyl α-L-fucoside	0.1	0.1	0.3	0.07
p-Nitrophenyl β-L-fucoside	0.06	0.06	0.3	0.07
L-galactose	0.2	0.4	0.6	1.2
D-galactose	N	N	N	N
D-mannose	0.2	0.2	0.3	0.3
L-mannose	N	N	N	N
p-Nitrophenyl α-D-mannoside	0.1	0.1	3.0	0.7
Methyl α-D-mannoside	0.09	0.2	0.5	0.5
D-mannoheptulose	0.3	0.7	0.5	1.0
D-mannosamine	N	N	N	N
D-mannose-6-phosphate	0.7	0.3	0.4	0.4
Yeast mannan	0.01%	0.01%	0.02%	0.02%
D-glucose	N	N	N	N
D-altrose	6.0	6.0	5.0	5.0
D-talose	N	N	N	N

[a] Determined with bacterial suspensions 10 times the minimal coagglutinating dose
[b] N, no inhibition at 17 mM final concentration

structural relatives of these sugars to inhibit coagglutination. Table 3 shows the quantitative similarity between the coagglutination inhibitions obtained with *A. salmonicida* 394 and 436 and human blood cells or yeast cells. These results provide additional information on the structural characteristics recognized by the adhesin. Figure 1 shows that the spatial arrangement of the hydroxyl groups on carbon atoms 2, 3, and 4 of D-mannose relative to the pyranose ring is identical to the arrangement of hydroxyls on C-4, C-3, and C-2 of L-fucose if the molecules are viewed with the C-1 of D-mannose and the C-6 of L-fucose on the right. These three hydroxyl groups are probably essential for adhesin recognition. This is confirmed by the observation that D-glucose, which differs from D-mannose at C-2, and D-mannosamine do not inhibit. Similarly D-talose and D-altrose, which differ from D-mannose at C-4 and C-3 respectively, are either not inhibitory or are poorly inhibitory. D-galactose, which differs at C-2 and C-4, also fails to inhibit. Furthermore coagglutination is inhibited by yeast mannan, which is a polymer of D-mannose in which the terminal sugar groups are almost exclusively D-mannose-linked to the rest of the polymer through C-1, with the hydroxyls at C-2, C-3, and C-4 available for adhesin recognition [3]. Both mannosides and fucosides inhibit the adhesin. Indeed both α- and β-linked fucosides are equally effective in their ability to inhibit, and so the adhesin of *A. salmonicida* is apparently able to re-

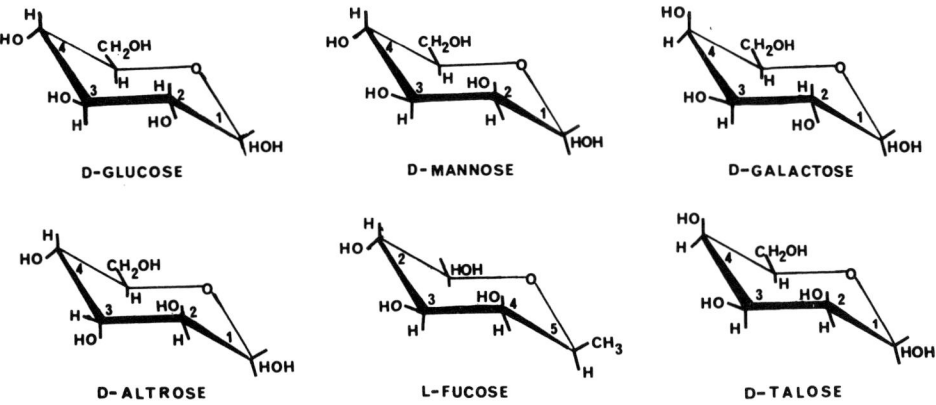

Fig. 1. Structures of D-glucose, D-mannose, D-galactose, D-altrose, L-fucose, and D-talose

cognize either α- or β-linked terminal molecules of either L-fucose of mannose. The inhibition pattern exhibited by this *A. salmonicida* adhesin is similar to that exhibited by the pilus adhesin of *A. hydrophila* 6 (to be published). However, the pilus adhesin is not inhibited by D-altrose, and its primary recognition structure is probably an α-linked L-fucose.

In summary, this study has shown that species of *Aeromonas* have adhesins which allow them to attach to a variety of eukaryotic cells, including trout blood cells. The adhesin of *A. salmonicida* recognizes D-mannose of L-fucose either α- or β-linked to polymers. This adhesin may be important in determining the organism's ability to attach to fish epithelial cells, so may have a key role in the virulence of *A. salmonicida*.

Acknowledgments. This work was supported by Grants A6232 and G0246 from the Natural Sciences and Engineering Research Council of Canada.

References

Duguid JP (1959) Fimbriae and adhesive properties in *Klebsiella* strains. J Gen Microbiol 21: 271–286
Jones GW (1977) The attachment of bacteria to the surfaces of animal cells. In: Reissig JL (ed) Receptors and recognition, ser B, vol III. Microbial interactions. Chapman and Hall, London p 139
Nakajima T, Ballou CE (1974) Characterization of carbohydrate fragments obtained from *Saccharomyces cerevisiae* mannan by alkaline degradation. J Biol Chem 249:7679–7684
Old CD (1972) Inhibition of the interaction between fimbrial hemagglutinins and erythrocytes by D-mannose and other carbohydrates. J Gen Microbiol 71:149–157

Bacterial Stress-Caused Infections of Silver Carp and Sarotherodon aureus in Fish Ponds and Their Control

S. SARIG and I. BEJERANAO[1]

Abstract

During the transfer of *S. aureus* and silver carp from one pond to another in the fish farms in Israel in autumn and in summer, heavy fish mortalities occur sometimes up to 100%. Yearly losses reach more than 150 tons of fish, more than 5% of the total production.

The preventive methods in practice until now are a mixture of 25 ppm formalin and 0.15 ppm malachite green, or concentrations of 25 mg/1 active ingredient of furazolidone and oxytetracycline or chloramphenicol and oxytetracycline. In all cases the results were only partially positive. Also, no controls could be carried out.

There are several indications that these mortalities are mainly due to stress during hawling, sorting, and transportation of the fish. A secondary factors as a result of such stress is increased sensitivity of the fish to surrounding damaging factors, i.e., bacteria or fungi.

A model of artificially caused stress under laboratory conditions was developed in our laboratory in order to investigate these developments in more detail. Stress was obtained by reduction of temperature to about 15 °C, and producing "social" stress in the aquaria. It was easy to identify specific external signs of stress such as behavior, pigmentation, loss of scales, mucus, and hemorrhage as well as some hematological changes like in hematocrit, hemoglobin, lymphocytes, red blood cells, glucose, and total protein in plasma.

It was possible to obtain sublethal stress and under such conditions intraperitoneal injections with the bacteria *Proteus rettgeri* NCIB 9575 caused etiology exactly as in the ponds, terminating in almost total mortalities.

---

1 Laboratory for Research of Fish Diseases, Nir-David, 19150, Israel

Physiological and Morphological Effects of Social Stress on the Eel, Anguilla anguilla L.[1]

Gabriele PETERS, H. DELVENTHAL, and H. KLINGER[2]

Stress is defined as the strain placed on a vertebrate under extreme endogenic or exogenic stimulation. All nonspecific physiological reactions of the body to this encumbrance are designated as stress effects (Selye, 1956). Gronow (1974) proved that the concept of stress is transferable from mammals to fish.

We investigated the effects of social stress in the eel. Under conditions of captivity, eels develop aggressive behavior patterns consisting of threatening gestures and biting. This aggression is particularly pronounced when the stocking density is low. To find out what effects the establishment of a hierarchy will have on the fish, 56 eels were kept in pairs and 18 control eels were held isolated in 50-l aquaria which were part of a warm-water circulating system (22 °C). To judge the degree of strain, 20 physiological and morphological parameters were measured and recorded.

In every case, battles to determine rank were fought between the eels during the five- to ten-day experimental period. These occasionally ended in the death of the subordinate eel(β). The dominant eel(α) resemble the controls in almost all parameters. Subordinate animals show numerous statistically significant changes. The removable blood quantity and spleen weight decrease. The plasma cortisol concentration increases in most subordinate eels, but in some β-eels it decreases. The cells of the interrenal tissue are enlarged and show a greater structural heterogeneity. In the subordinate eels a decrease in liver glycogen and an increase in blood glucose and lactate could be observed. In β-eels changes in white blood composition occur too. Their total leukocyte counts are lower mainly due to a decrease in lymphocytes, however their leukocrits increase. This is due to the greater number of large granulocytes. Hemoglobin appears in the skin mucus of β-eels. In general the physiological parameters of the inferior fish show a higher variability (Table 1).

The symptoms observed in the subordinate eels are mostly characteristic of the General Adaptation Syndrome (GAS) of Selye's stress principle. The hypothalamus-hypophysis-adrenal cortex system and the hypothalamus-sympathetic nerve-adrenal medulla system are known to be the two superior pathways of the stress reaction. In fish the inter- and suprarenal glands are homologous to cortex and medulla of the adrenal gland in higher vertebrates. Cortisol as the main glucocorticoid in fish is produced in the interrenal gland. It influences in connection with the catecholamines the

---

[1] The full paper will be published in *Archiv für Fischereiwissenschaft* 30, 2–3 (1980). Some results of this publication are part of the theses of H. Delventhal and H. Klinger.

[2] Institut für Hydrobiologie und Fischereiwissenschaft, Universität Hamburg, Olbersweg 24, D-2000 Hamburg 50, FRG

Table 1. Summary of the data from the social stress experiments

Investigated Parameters	Control N	x̄	sD	α N	x̄	sD	β N	x̄	sD
Length (cm)	18	42.7	± 3.9	28	43.5	± 4.9	28	39.7	± 4.7
Weight (g)	18	102.9	± 28.2	28	109.4	± 38.4	28	81.7	± 28.6
Gutted weight (g)	18	94.4	± 26.7	28	99.7	± 35.5	28	75.2	± 27.0
K-factor	18	0.131±	0.025	28	0.127±	0.021	28	0.126±	0.021
Liver som. ind.	18	1.183±	0.178	28	1.419±	0.466	28	1.366±	0.394
Spleen som. ind	18	0.074±	0.025	28	0.088±	0.055	28	0.067±	0.027
Hematocrit (vol.%)	18	45.1	± 5.1	27	45.5	± 4.6	27	45.1	± 7.4
Leucorit (vol.%)	18	1.6	± 0.5	19	1.3	± 0.3	19	2.0	± 0.8
Erythrocyte count (x10⁶)	18	2.10	± 0.15	17	2.09	± 0.26	23	2.15	± 0.30
Leucocyte count (x10³)	17	120.6	± 31.4	17	103.2	± 31.3	23	81.9	± 22.7
Hemoglobin content (g%)	18	13.8	± 1.9	27	14.1	± 4.0	27	14.5	± 3.7
Blood glucose (mmol/1)	18	5.72	± 1.68	19	5.53	± 1.61	19	13.42	± 8.51
Blood lactate (mmol/1)	18	0.42	± 0.39	19	0.62	± 0.38	16	1.68	± 1.15
Liver glyc. un. (µmol/g)	18	385.4	±207.8	28	401.3	±201.2	28	167.0	±155.6
Plasma cortisol (nmol/l)	18	169.9	± 70.6	23	157.3	± 91.0	21	227.8	±181.4
Interren. nucl. ∅ (µm)	4	4.73	± 0.48	23	4.60	± 0.16	18	4.48	± 0.45
Interren. nucl. area	4	17.14	± 0.92	18	17.38	± 3.91	16	12.03	± 4.60
Total plasma protein (g%)	17	5.2	± 0.76	26	4.67	± 1.00	24	5.55	± 1.36
Tot. pl. cholest. (mg/100 ml)	18	518.8	±132.9	23	460.7	±113.6	17	455.6	±144.2
Blood vol. obt. (10⁻⁶ ml/g)	17	1.18	± 0.37	24	1.3	± 0.32	24	0.84	± 0.34
Blood smear:	15			22			20		
Thrombocytes		59.000			47.000			38.000	
Lymphocytes		41.000			43.000			14.000	
Granulocytes		12.000			12.000			28.000	

carbohydrate metabolism in order to support a sufficient energy supply in stress situations. It is known in mammals that the glucocorticoids show an immunosuppressive effect.

Thus the establishment of a hierarchy in captivity results in social stress with clinically measurable effects, which are often irreversible.

Acknowledgment. Supported by the Bundesministerium für Forschung und Technologie, grant number MFE 0329/5.

References

Gronow G (1974) Über die Anwendung des an Säugetieren erarbeiteten Begriffes "Stress" auf Knochenfische. Zool Anz 192:316–331
Selye H (1956) General Physiology and pathology of stress. V. Report on stress. MD Publ Inc, New York

Epitheliocystis Disease in Fishes

I. PAPERNA and I. SABNAI[1]

Introduction

Epitheliocystis was first described by Hoffman et al. (1969) as a benign chronic infection of fishes. In recent years it became evident that epitheliocystis infections in fishes may be very pathogenic and cause severe mortalities to fish in culture, particularly to juvenile fish in nurseries. Epitheliocystis infections are widely distributed and occur in many species of freshwater as well as marine fish.

Cause

Intracellular infection of the skin and gill epithelial cells and the gill's chloride cells by prokaryotic organisms, so-called epitheliocystis organisms.

Gross Signs

Infection results in extreme hypertrophy of the infected cells, leading to the formation of distinct white transparent capsules of variable sizes (35–100x55–55 μm) on the gill filaments or within the skin epithelium.

In severe infections the gills may lose their lamellar structure, swell, and adhere to one another.

Fish Species Affected and Known Geographic Distribution

Epitheliocystis infections occur in freshwater as well as in marine fish. To date, infections have been found in the blue gill (*Lepomis macrochirus;* Hoffman et al., 1969), the striped bass *(Morone saxatilis)*, and the white perch *(Morone americanus)* (Wolke et al., 1970; Zachary and Paperna, 1977) in the eastern United States; in *Tilapia mossambica* and *Tilapia nilotica* in South Africa and Israel; in carp *(Cyprinus carpio)* in Israel; in several species of mullets (Mugilidae: *Mugil cephalus, Liza ramada, Liza aurata, and Liza subviridis*) in the east Mediterranean Sea and in the Gulf of Aqaba,

---

[1] H. Steinitz Marine Biology Laboratory, Eilat, and the Zoology Department of the Hebrew University of Jerusalem, Israel

Red Sea; in *Upeneus mollucensis* from the Mediterranean cost of Israel; in the sea bream *(Sparus aurata)* from the east Mediterranean region (Bardawil Lagoon, north Sinai) and from the Mediterranean coast of France, and in the sea bass *(Dicentrarchus labrax)* from the Mediterranean and Atlantic (Brittany) coasts of France (Paperna, 1977; Paperna et al., 1978; Paperna and Baudin-Laurencin, 1979).

Diagnosis

The intracellular epitheliocystis organisms are pleomorphic, either rod-shaped and 1–2 μm in length or coccoids 0.3–0.5 μm in diameter, and therefore are best demonstrated by transmission electron microscopy in ultrathin sections made from the capsules. Light-microscope examination of lightly compressed fresh unstained gill filaments or skin scrapings, when infection is present, reveals transparent cysts with homogeneous granular substance. In hematoxylin-eosinstained histological preparations, in early-stage infection epitheliocysts organisms occur as a single inclusion with granular basophilic contents within the infected cell. As cellular hypertrophy progresses in an advanced stage of infection, epitheliocystis-infected cells develop into large capsules or cysts with a distinct eosinophilic cell wall and basophilic granular content, lined by a layer of epithelial cells.

Ultrastructural Characteristics

Electron micrographs of epitheliocystis organisms are so far available from infections in blue gill, striped bass (Hoffman et al., 1969; Wolke et al., 1970; Zachary and Paperna, 1977), sea bream, grey mullets (Paperna et al., 1978), and sea bass. In the micrographs from all the above-mentioned fish hosts the organisms show similar characteristics. They are bounded by outer and inner trilaminated membranes similar to those seen in rickettsiae and chlamydiae and the cytoplasm containing aggregates of ribosome-like particles interspersed with fine fibrils (presumably DNA). In most cells a dense nucleoid area is present at the cell center. A progressive binary fission process is evident in many cells. Electron micrographs of epitheliocystis organisms from the striped bass revealed rod-shaped rickettsia-like organisms. The outer membrane of many cells near the capsule periphery appeared rippled and separated from the inner membrane and in many cells resulted in the formation of large sac-like structures (Zachary and Paperna, 1978). A similar phenomenon of membrane separation was also seen in electron micrographs from the blue gill (Hoffman et al., 1969).

Electron micrographs prepared from epitheliocystis cysts from sea bream and grey mullets revealed the existence of at least three distinct morphological forms: rounded cells (RC), elongated cells (EC), and small cells (SC). All three forms occur in electron micrographs from infected gills of sea bream as well as from grey mullets. The RC (0.36–0.58 μm in diameter) contain a distinct electron-dense nucleoid surrounded by electron-transparent cytoplasma, following binary fission the RC remain interconnected by plasmatic bridges (Fig. 1).

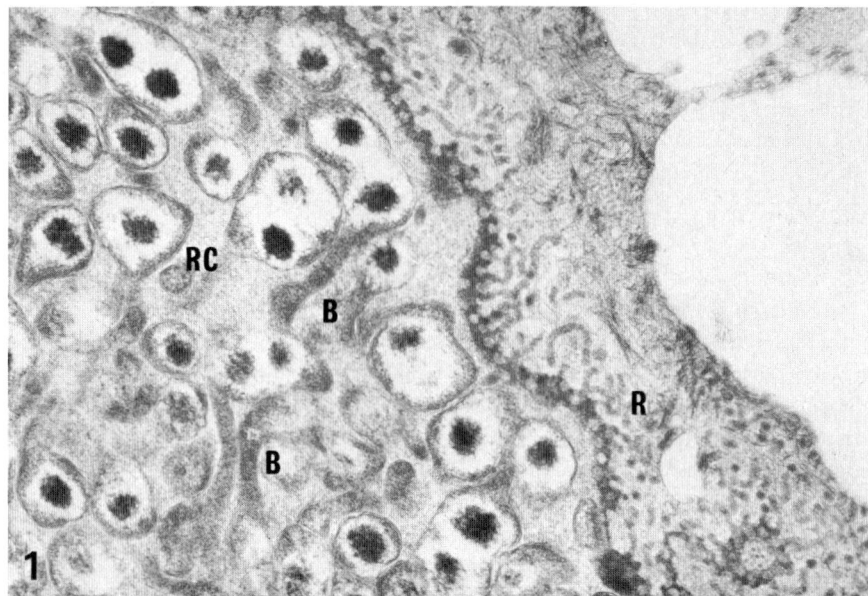

Fig. 1. Round cells *(RC)* within infected chloride cells from gills of *M. cephalus*. B plasmatic bridges; R tubular reticulum of the host cell cytoplasm. TEM x 27,000

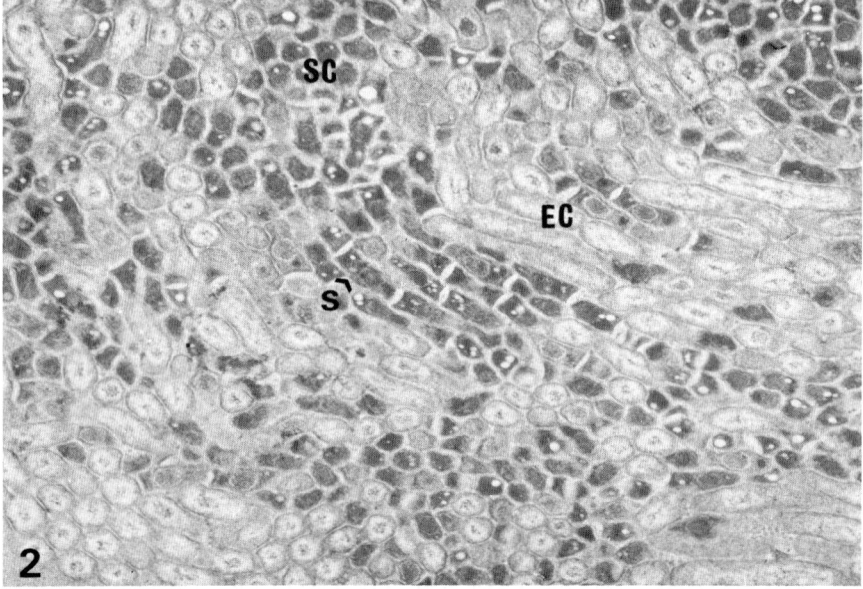

Fig. 2. Elongated cells *(EC)* and smal cells *(SC)* seen within epitheliocystis-infected epithelial cells from gills of *M. cephalus*. S zone of multiple division of *EC* to *SC*. TEM x 12,600

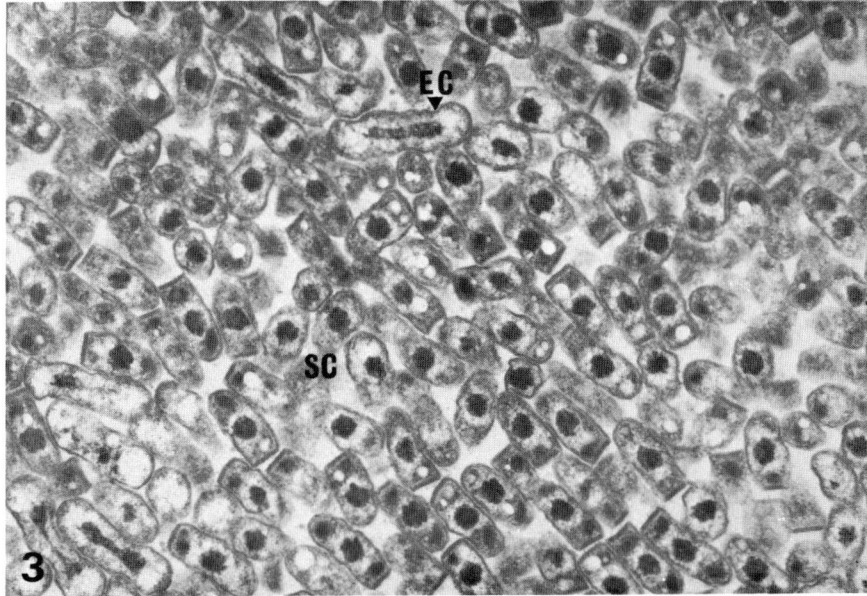

Fig. 3. Elongated cells *(EC)* and mature small cells *(SC)* within infected epithelial cells from gills of *M. cephalus*. TEM x 18,000

The ECs, rod-shaped, 1.9–5.4x0.3–0.5 μm in size, with a less pronounced nucleoid, were observed either dividing by binary fission into two daughter EC, or through simultaneous multiple fission producing SC (Fig. 2, 4). The SC from sea bream are rounded (0.46–0.48 μm in diameter), while in grey mullets they are peg-shaped, or coccoids (0.60–0.83x0.26–0.43 μm in size) with one rectangular end (Fig. 3, 4). Both forms have electron-dense cytoplasm and contain one or more vacuoles.

RC were observed in both sea bream and grey mullets only within the epithelial chloride cells of the gills, recognised by the presence of extensive tubular reticulum in the residual cell host cytoplasm around the epitheliocystis inclusion. All other stages were observed only in infected epithelial cells, or in host cells in which the cytoplasm was completely replaced by the epitheliocystis inclusion. Epitheliocystis organisms were originally thought to be related to chlamydiae (Hoffman et al., 1969; Wolke et al., 1970). By their morphology RC are reminiscent of chlamydiae. Moreover, the finding of several morphological forms of which at least two, EC and SC, comprise successive developmental stages, suggests a developmental cycle analogous with that characteristic of chlamydiae. On the other hand EC are reminiscent of rickettsiae, while the plasmatic bridges connecting the post-dividing RC stage and the simultaneous multiple division of EC to SC are unique features unknown among rickettsiae or chlamydiae of homeothermic hosts.

However, some remote relationship might be suggested between epitheliocystis organisms and Rickettsiella (Chlamydiales) (Morel, 1977; Louis et al., 1977) parasitic in Arthropodes. The "initial bodies" closely resemble EC of epitheliocystis, and in a

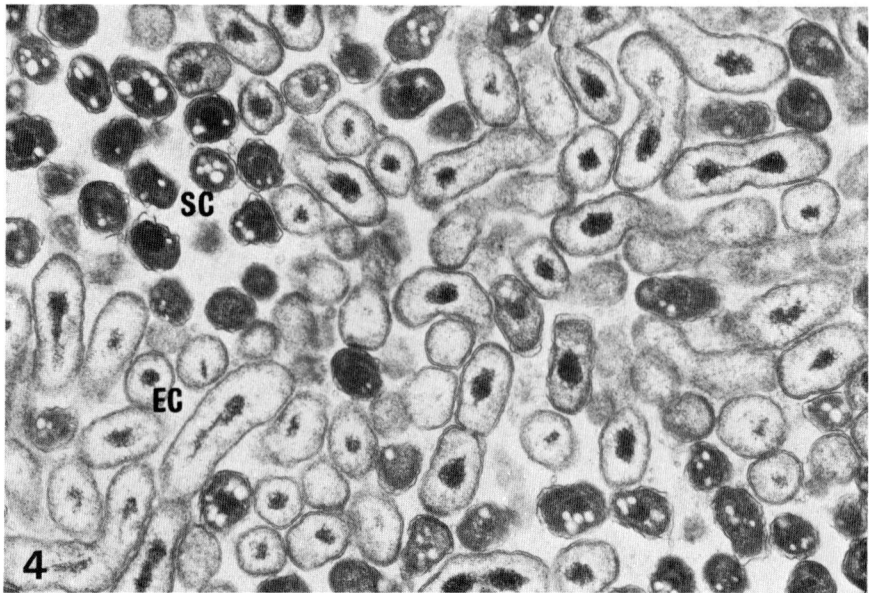

Fig. 4. Elongated cells *(EC)* and small cells *(SC)* within infected epithelial cells from gills of *S. aurata*. TEM x 18,000

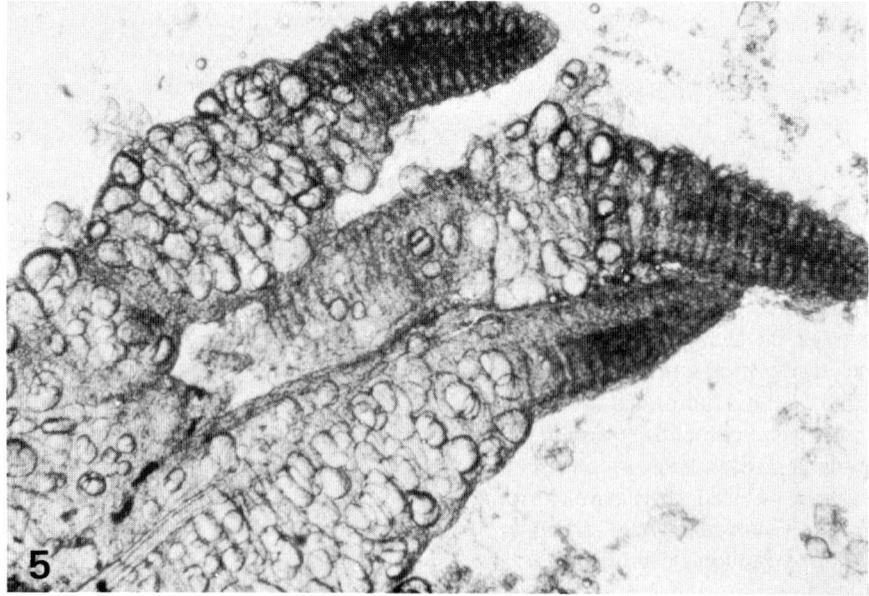

Fig. 5. Heavy infection with epitheliocystis proliferative condition in gills of *S. aurata*, unstained. x 100

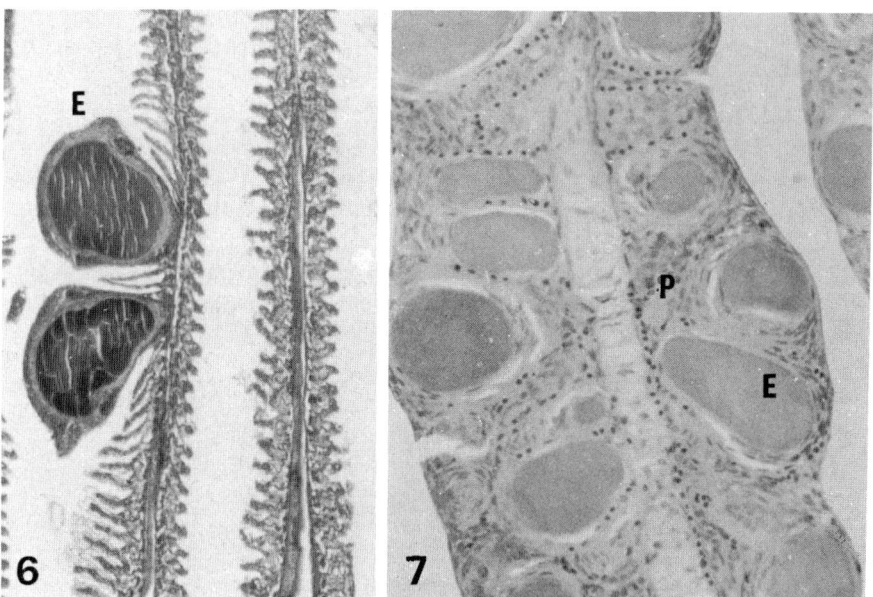

Fig. 6. Cysts of epitheliocystis *(E)* in gills of *Morone saxatilis* benign infection. HE-stained histological section. x 100

Fig. 7. Proliferative epitheliocystis infection of *S. aurata* gills. Epitheliocystis cysts *(E)* are embedded in proliferative epithelial tissue *(P)*. HE-stained histological section, x 450

few of these prokaryots, as in epitheliocystis multiple division of initial bodies into "elementary bodies" (analogous to SC of epitheliocystis), have been reported.

Epizootology and Pathology

Epitheliocystis occurs in fish either as a "benign infection" when infection in the gills induces only limited tissue response, or as a "proliferative condition" accompanied by extreme epithelial hyperplasia (Fig. 5). In benign infection tissue changes are limited to the formation of a thin epithelial enveloping layer around the developing capsule, sometimes also accompanied by infiltration of blood lacunae into the capsule which extends from the respiratory capillary network (Fig. 6). In the proliferative condition hyperplasia of the gill epithelium will result in the complete embedding of the respiratory capillary system by the proliferating epithelial with consequent impairment of the respiratory function of the gills (Fig. 7).

Benign infection occurs as a chronic nonpathogenic condition in fish of all ages, in natural habitats as well as in culture conditions. Benign infection will transform into a proliferative condition which will consequently cause severe mortalities when young fish are maintained in adverse culture conditions. No recurrent proliferative infection has ever been observed in fish surviving epitheliocystis epizootics. So far proliferative infections with consequent outbreak of mortalities have occurred in sea bream and grey mullets *(mugil cephalus)* cultured in sea-water tanks in Eilat (Gulf of Aqaba, Red Sea) and in carp kept in freshwater tanks in Nir David, Israel (Paperna, 1977). Mortality has been reported also in cultured blue gills (in the United States) from skin as well as gill infection by epitheliocystis (Hoffman et al., 1969; K. Wolf, personal communication, 1979). Mild to severe epitheliocystis infection not accompanied by evident outbreak of mortality was observed in the following species of cultured fish: *Tilapia mossambica* (South Africa, Israel); *Tilapia nolitica* (Israel); sea bream (Séte, France) and sea bass (Eilat, Gulf of Aqaba, Séte and Brest, France) (Paperna and Baudin-Laurencin, 1979).

Acknowledgments. This research was supported by a grant from the GKSS Geesthacht-Tesperhude, Germany.

References

Hoffman GL, Dunbar CE, Wolf K, Zwillenberg LO (1969) Epitheliocystis, a new infectious disease of the bluegill *(Lepomis macrochirus).* Antonie van Leeuwenhoek. J Microbiol Serol 35: 146–158

Louis C, Yousfi A, Vago C, Nicolas G (1977) Etude par cytochimie et cryodécapage de l'ultrastructure d'une *Rickettsiella* de crustacé. Ann Microbiol Inst Pasteur 128B:177–205

Morel G (1977) Etude d'une *Rickettsiella* (Rickettsie) se developpant chez un Arachnidae, l'araignée *Pisaura mirabilis.* Ann Microbiol Inst Pasteur 128A:49–59

Paperna I (1977) Epitheliocystis infections in wild and cultured sea bream *(Sparus aurata,* Sparidae) and grey mullets *(Liza ramada,* Mugilidae). Aquaculture 10:169–176

Paperna I, Baudin-Laurencin F (1979) Parasitic infections of sea bass, *Dicentrarchus labrax* and gilt-head bream, *Sparus aurata,* in mariculture facilities in France, Aquaculture 16:173–175

Paperna I, Sabnai I, Castel M (1978) Ultrastructural study of epitheliocystis organisms from gill epithelium of the fish *Sparus aurata* (L.) and *Liza ramada* (Risso) and their relation to the host cell. J Fish Dis 1:181–189

Wolke RE, Wyand DS, Khairallah LH (1970) A light and electron microscopic study of epitheliocystis disease in the gills of Connecticut striped bass *(Morone saxatilis)* and white perch *(Morone americanus).* J Comp Pathol 80:559–563

Zachary A, Paperna I (1977) Epithelicystis disease in the striped bass *Morone saxatilis* from the Chesapeake Bay. Can J Microbiol 28(10):1404–1414

Contributions to the Taxonomy of the Genus Diplozoon von Nordmann, 1832

H.-H. REICHENBACH-KLINKE[1]

Summary

An attempt is made to sum up the differences between the nearly 30 species of this genus in order to find exact characters which allow the introduction of a valid table. The specifications are based on the characterization of the development, the morphology, and the chromosome picture.

The species of the genus *Diplozoon* von Nordmann (1832) the so-called "double animalcules", live in the adult stage in the gills of freshwater fishes, without extensive pathogenicity combined with fish losses. Diplozoon is defined as a genus of the class Monogenea, order Diplozoidea Yamaguti (1963) family Diplozoidae Yamaguti (1963) in which two adult specimens are in every case united in x-form. With the new description of the species *Neodiplozoon barbi* Tripathi (1959), we found another genus with this characterization. This genus *Neodiplozoon* differs from *Diplozoon* by the high number of clamps, mostly 16, in this point similar to the Microcotylidae. The definition of the genus *Diplozoon* therefore must include the sign: 4 clamps on both sides of the posterior part, then 1 pair of little hooks (hamuli) at the end of the body.

The number of the species described to date is more then 20. In a monography of the genus *Diplozoon* the author has tried to identify some of these species as subspecies, other ones as synonyma. This seems to be necessary, because one of the species characters, i.e., number and size of the clamps, varies significantly (Oliver and Reichenbach-Klinke, 1973). Bychowsky and Nagibina (1959) differentiated the number of gut diverticles in the posterior part of the body, Korolewa (1968) has determined the number of chromosomes. With these results a new system is arranged, demonstrated in the following paper.

After our investigations it seems to be evident that *D. homoion* Bychowsky and Nagibina is predominant in Central Europe. This species is split up into the subspecies *homoion, sapae, bliccae, gracile,* and probably others. The creation of subspecies is visible at the frontiers of the distribution area, where for instance in South-West France specimens with irregularities in the clamps are found. More attention should be directed to the number of chromosomes (Korolewa). It must be summarized that the genus *Diplozoon* is in a phase of activation and development. It is therefore necessary that taxonomic studies should be carried out all over the area covered.

---

[1] Institut für Zoologie und Hydrobiologie, Universität München, Kaulbachstraße 37, D-8000 München

Key to European Diplozoon

1 (14) Posterior part of the body enlarging smoothly to the end

2 (13) Gut in the posterior body with coeca (diverticles)

3 (10) Gut in the hind body with more than 5 diverticles

4 (9) Gut in the hind body poorly anastomosing, clamps with different measures

5 (8) Last clamp smaller than the first ones

6 (7) Last clamp two-thirds as broad as the last but one, shaft of the larval hamulus more than 60 μm *D. megan*

7 (6) Last clamp maximally half as broad as the last but one, shaft of the larval hamulus not more than 57 μm *D. nagibinae*

8 (5) Last clamp slightly smaller than the first ones, large larval hamuli (shaft up to 66 μm, hooks up to 30 μm) *D. rutili*

9 (4) Gut in the hind body anastomosing like a net, clamps nearly equal in size *D. paradoxum*

10 (3) Gut in the hind body with 5 or less diverticles, last clamps slightly smaller than the first ones

11 (12) Handle-ends of the clamps melt together, central buckle united with the front handle by a ribbon *D. markewitschi* (syn. *gussevi*)

12 (11) Handle-ends of the clamps not united, central buckle smoothly bent and grooved in the centre *D. homoion*

13 (2) Gut in the posterior body without or with short diverticles *D. pavlovskii*

14 (1) Posterior part of the body in the center enlarged belly-like on both sides, gut in the hind body without diverticles *D. nipponicum* (perhaps introduced?)

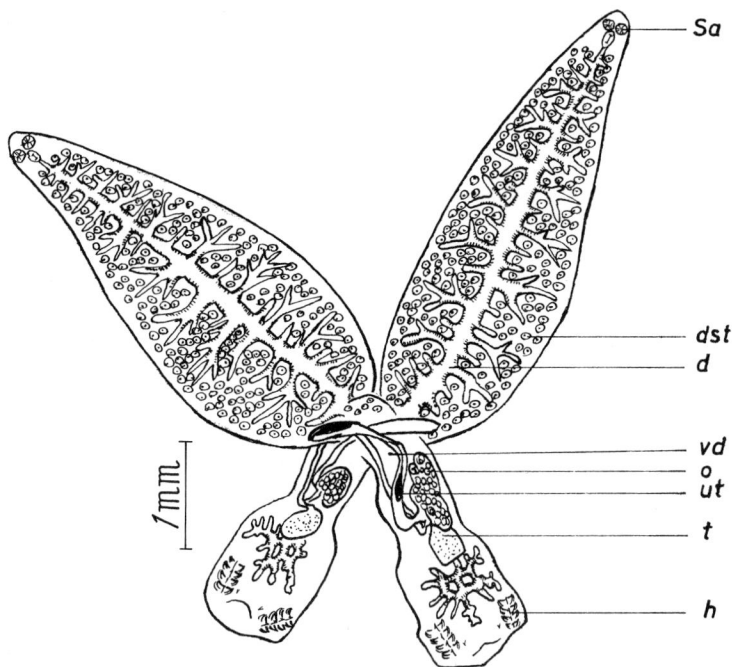

Fig. 1. *Diplozoon paradoxum* von Nordmann. *d* gut, *dst* yolk center, *h* hook, *o* ovary, *Sa* sucker, *t* testis, *ut* uterus, *vd* vas deferens

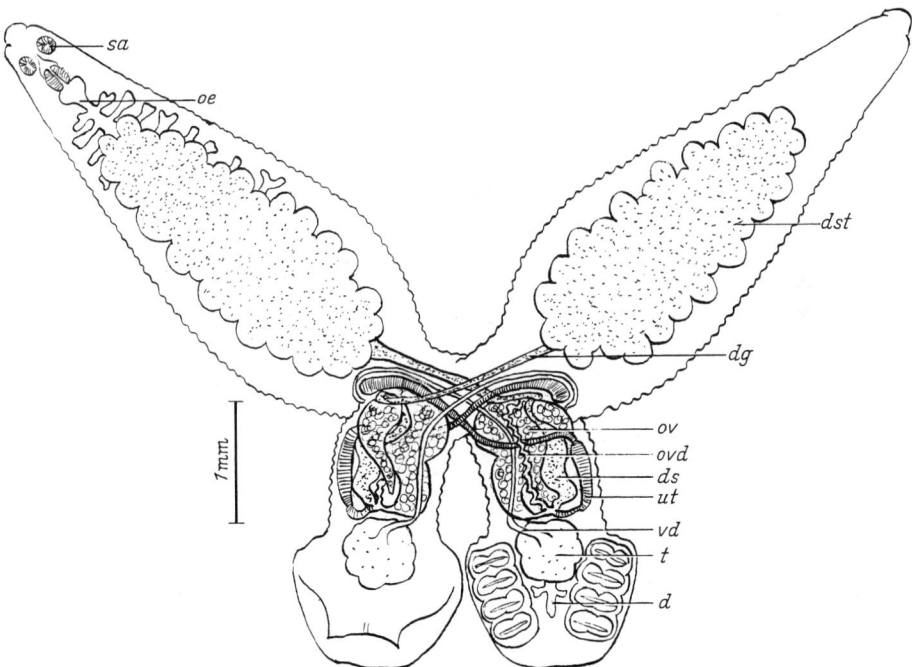

Fig. 2. *Diplozoon paradoxum* von Nordmann. *d* end of gut; *dg* yolk channel; *ds* yolk sac; *dst* yolk center; *oe* esophagus; *ov* ovary; *ovd* oviduct; *sa* sucker; *t* testis; *ut* uterus; *vd* vas deferens. (After von Nordmann, 1832)

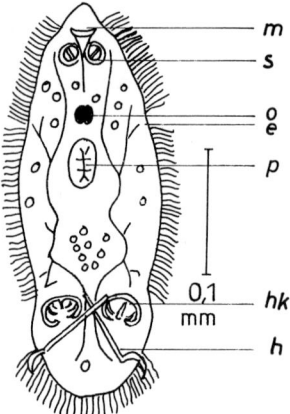

Fig. 3. Larval stage of *Diplozoon paradoxum* von Nordmann. *e* excretion pore; *h* larval hooks; *hk* clamp; *m* mouth; *o* ocellus; *p* pharynx; *s* oral suckers. (After Bychowsky, 1957)

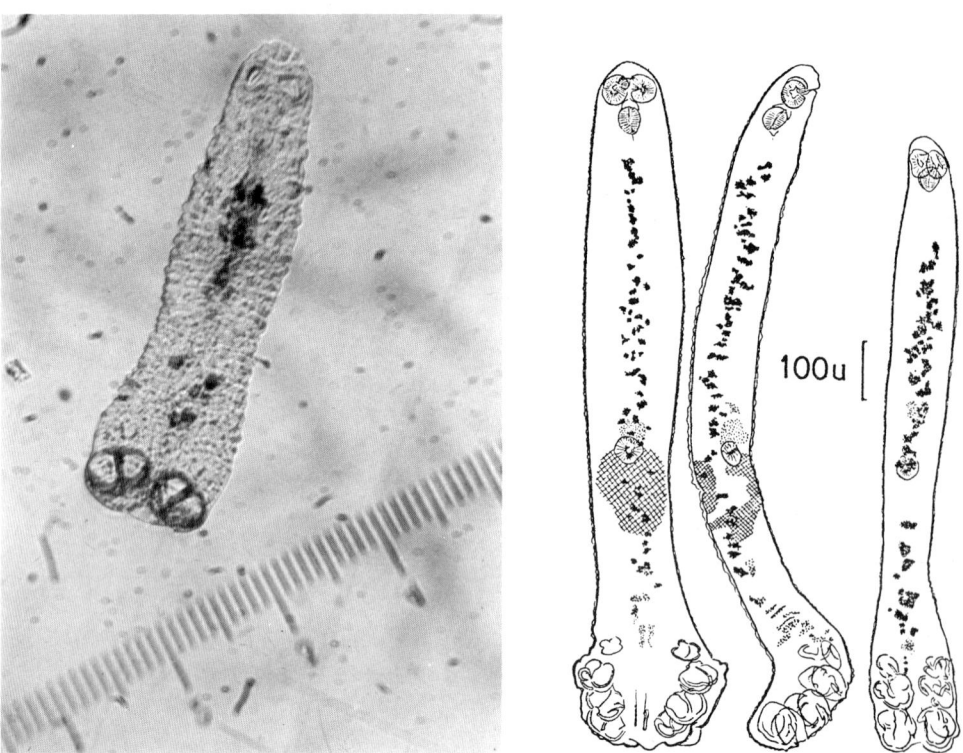

Fig. 4. Diporpa-Stage of *Diplozoon paradoxum* von Nordmann. *Middle* 4, *right* 3 pairs of clamps (ventral sucker visible). Nach Bovet. *Left* Diporpa with 2 clamps (Foto Smija)

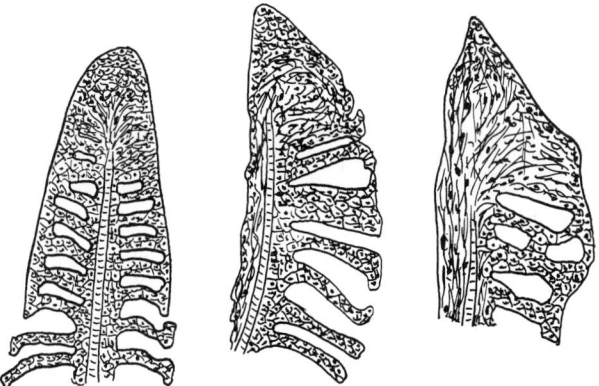

Fig. 5. Fish gills after parasitation by *Diplozoon*. The lamellae are swollen, hyperplastic, and grown together

Fig. 6

Fig. 7 *a* *b* *c* *d*

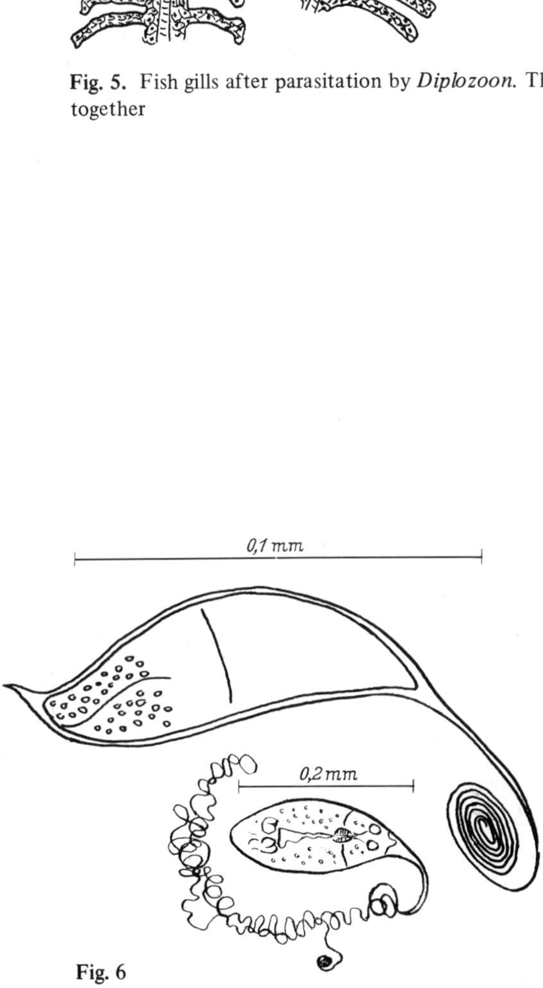

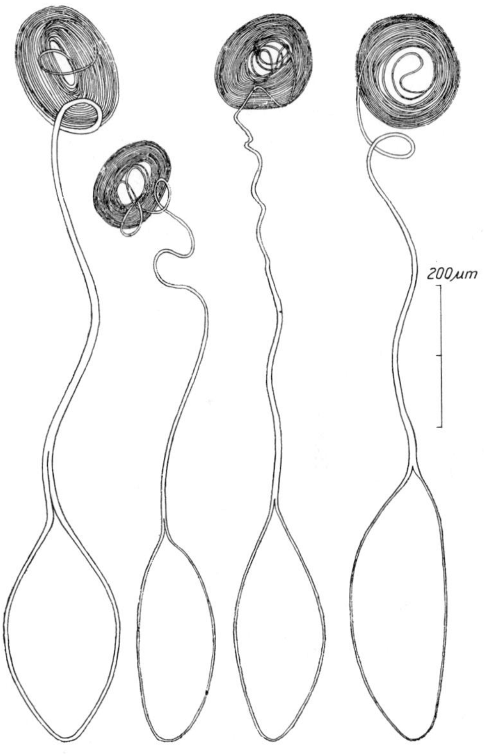

Fig. 6. Egg of *Diplozoon barbi* R.-Kl. *(left)* and *D. paradoxum* von Nordmann *(right)*

Fig. 7 a–d. Eggs of different *Diplozoon* species. **a** *D. rutili* Gläser; **b** *D. homoion* Bychowsky and Nagibina (B and N); **c** *D. nagibinae* B and N; **d** *D. paradoxum* von Nordmann. (After Gläser, 1967)

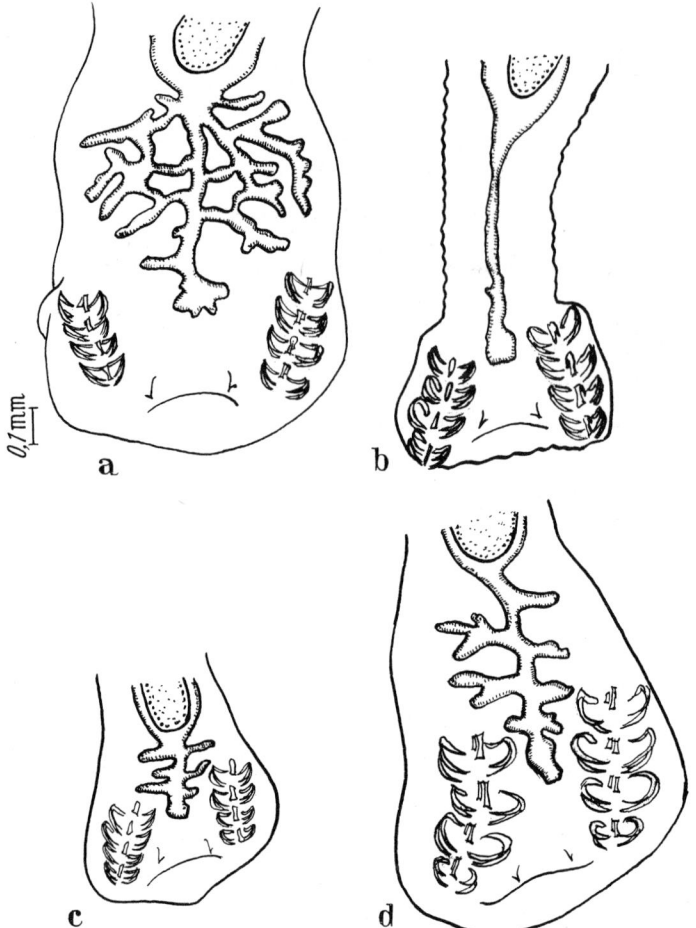

Fig. 8 a–d. Distal part of some *Diplozoon* species with the intestinal branches. **a** *D. paradoxum* von Nordmann; **b** *D. pavlovskii* B and N; **c** *D. homoion* B and N; **d** *D. megan* B and N. (After Bychowsky and Nagibina, 1959)

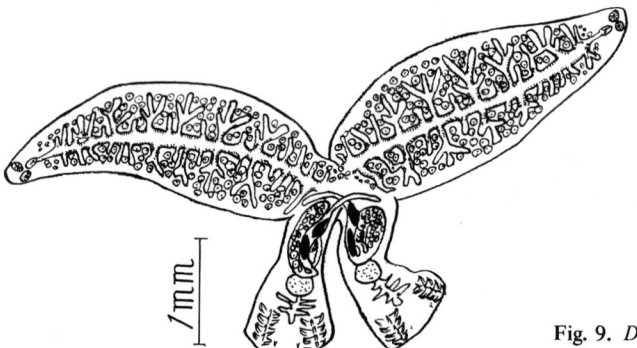

Fig. 9. *Diplozoon homoion* B and N. (After Bychowsky and Nagibina, 1959)

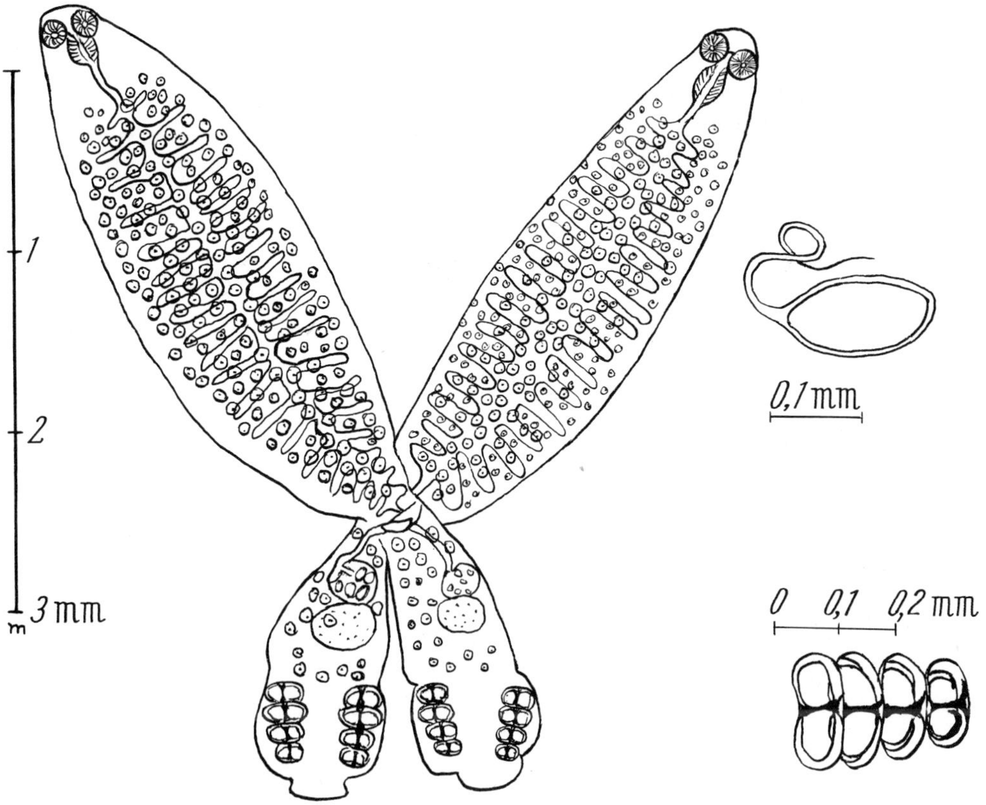

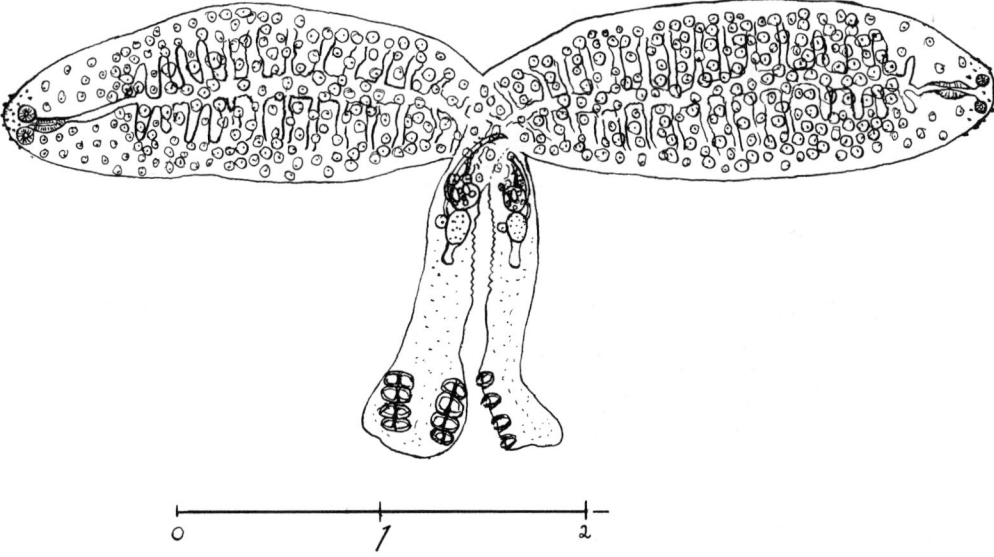

Fig. 10. *Diplozoon homoion sapae* R.-Kl.

Fig. 11. *Diplozoon homoion bliccae* R.-Kl.

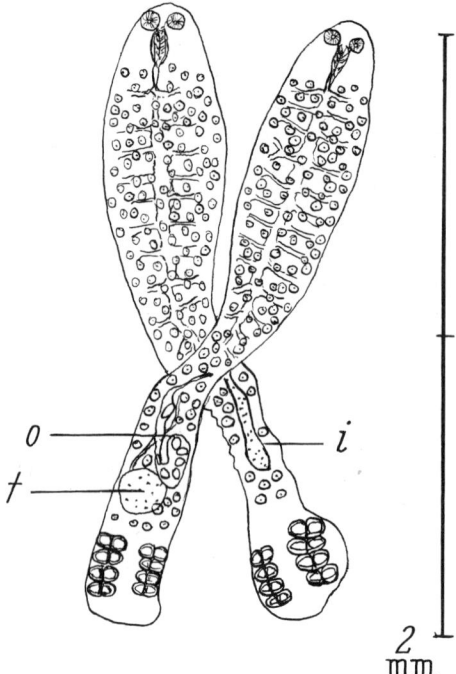

Fig. 12. *Diplozoon homoion gracile* R.-Kl. *i* intestine; *o* ovary; *t* testis

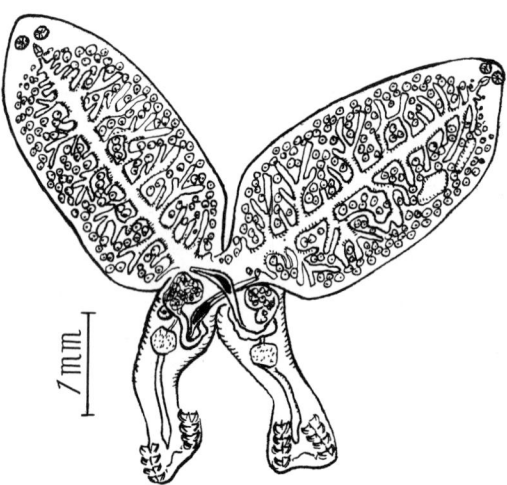

Fig. 13. *Diplozoon pavlovskii* B and N. (After Bychowsky and Nagibina, 1959)

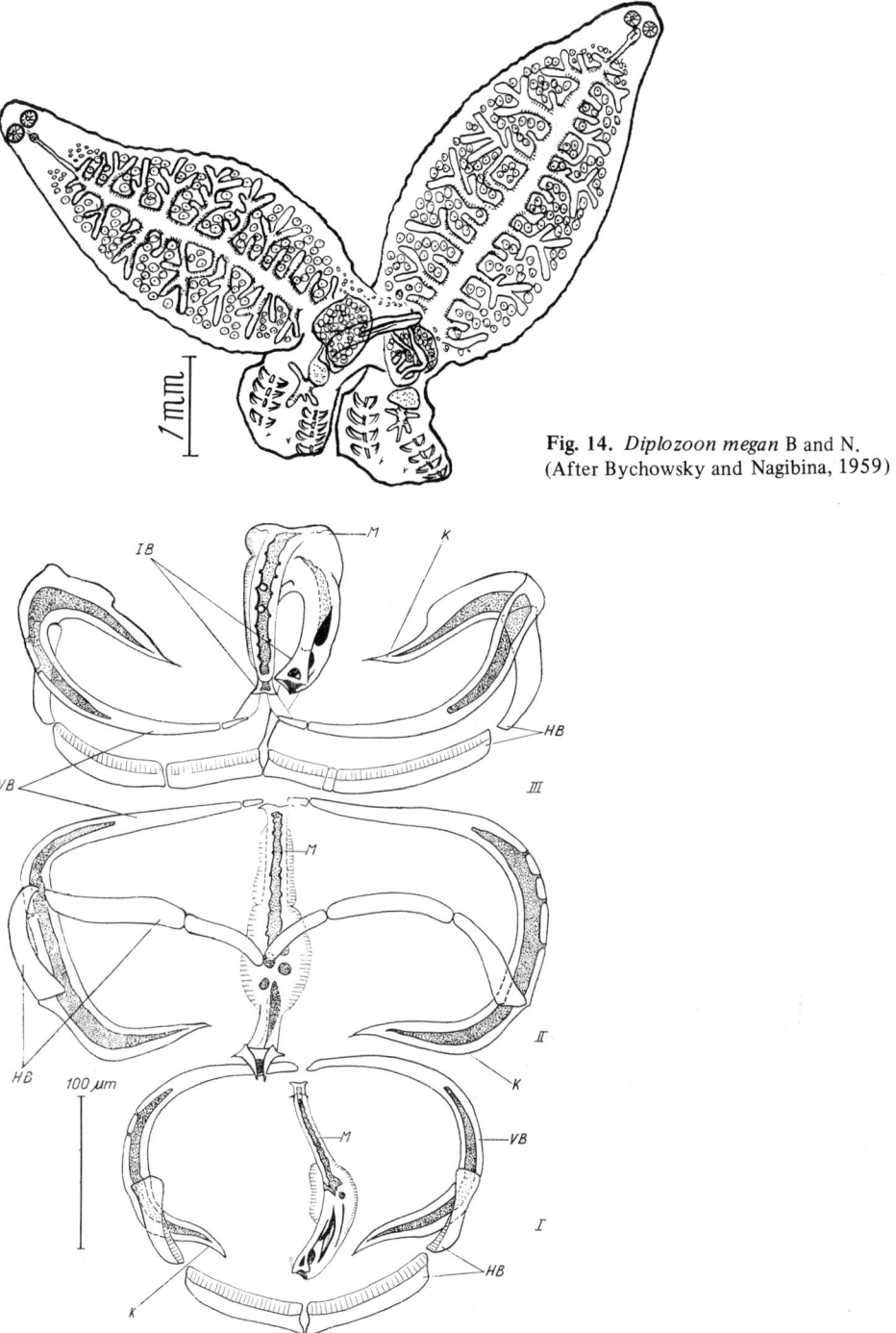

Fig. 14. *Diplozoon megan* B and N. (After Bychowsky and Nagibina, 1959)

Fig. 15. Sclerifizid parts of the clamps I–III of *Diplozoon rutili*. *HB* distal bend; *K* hock; *IB* inner bend; *VB* upper bend; *M* median bend. (After Gläser, 1967)

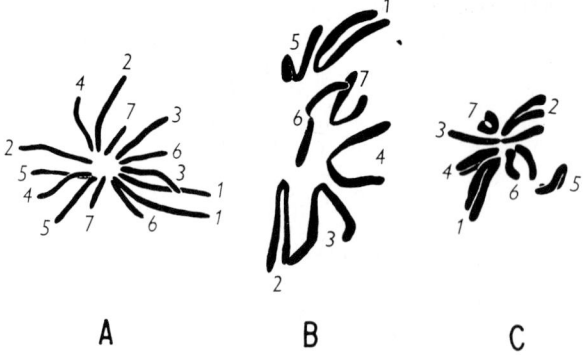

Fig. 16A–C. Chromosomes (diploid) of Diplozoon species **A** *Diplozoon paradoxum* Diporpa. Host: *Rutilus rutilus* N. Bovet. **B** *Diplozoon „paradoxum"* (?). Host: *Abramis brama* N. Bovet. **C** *Diplozoon „paradoxum"* (?). Host: *Rutilus rutilus* N. Bovet

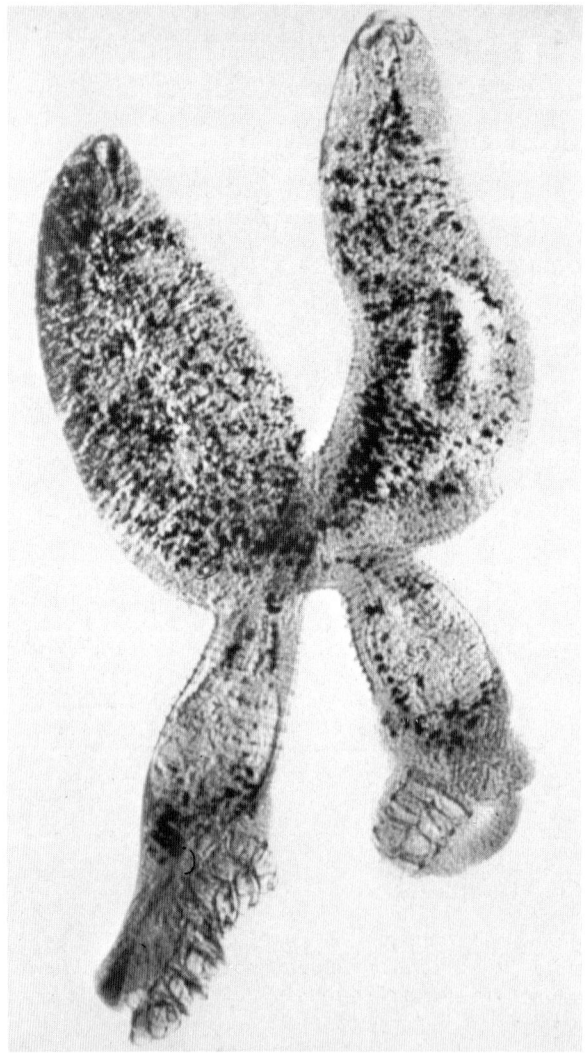

Fig. 17. *Diplozoon homoion gracile* R.-Kl.

Fig. 18. *Diplozoon nipponicum* Goto

References

Bovet J (1967) Contribution à la morphologie et à la biologie de Diplozoon paradoxum v. Nordmann, 1832. Bull Soc Neuchateloise Sci Nat 90:63–159
Bychowsky BE (1957) Monogenetic trematodes. Their systematics and phylogeny. (Russian English Translation Virginia Inst Mar Sci) Zool Inst Akad Nauk SSSR Moscow Leningrad, p 509
Bychowsky BE, Nagibina LF (1959) Über die Systematik der Gattung Diplozoon Nordmann. Zool Z 38:362–377
Gläser HJ (1967) Eine neue Diplozoon-Art (Plathelminthes, Monogenea) von den Kiemen der Plötze, Rutilus rutilus (L.). Zool Anz 178:333–342
Korolewa JI (1968) Karyologitscheskoje isutschenje nekotorych widow monogenei roda Diplozoon. Dokl Akad Nauk SSSR 179:739–741
Korolewa JI (1969) Karyologyj nekotorych widow Diplozoon. Parasitologija 3:411–413
Oliver G, Reichenbach-Klinke HH (1973) Observations sur le genre Diplozoon von Nordmann, 1832, en Languedoc-Roussillon. Ann Parasitol Hum Comp 48:447–456

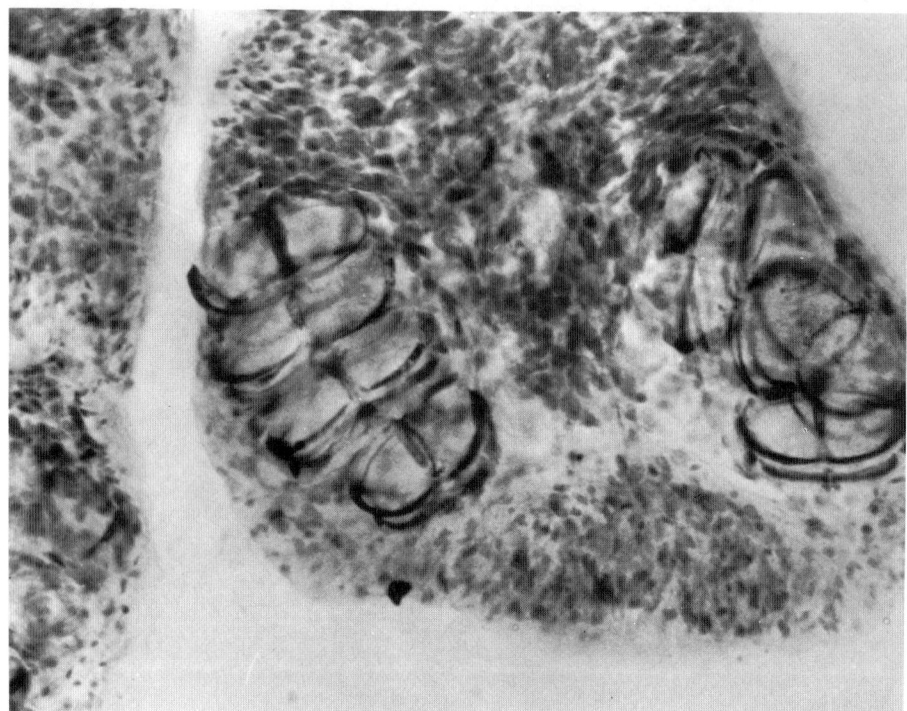

Fig. 19

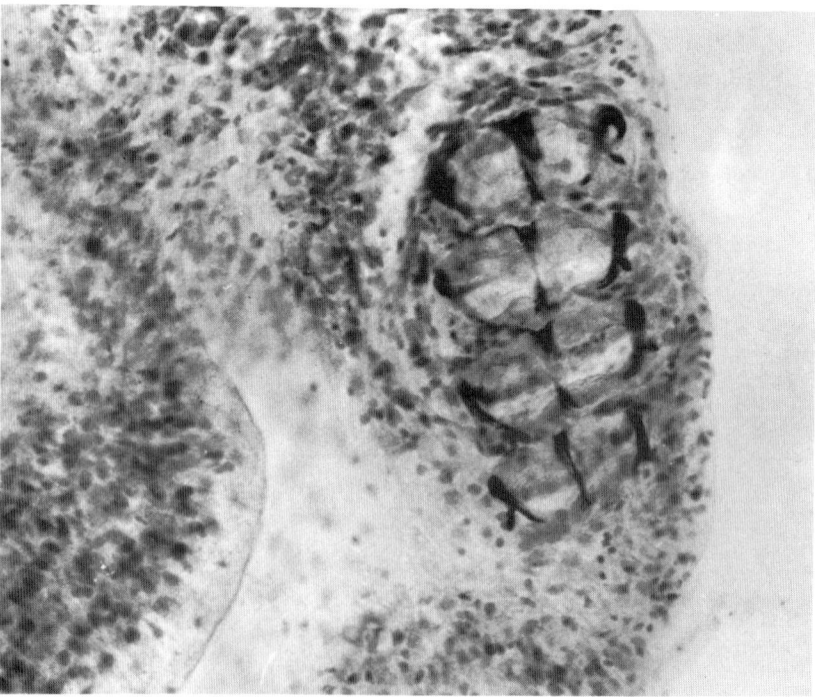

Fig. 20

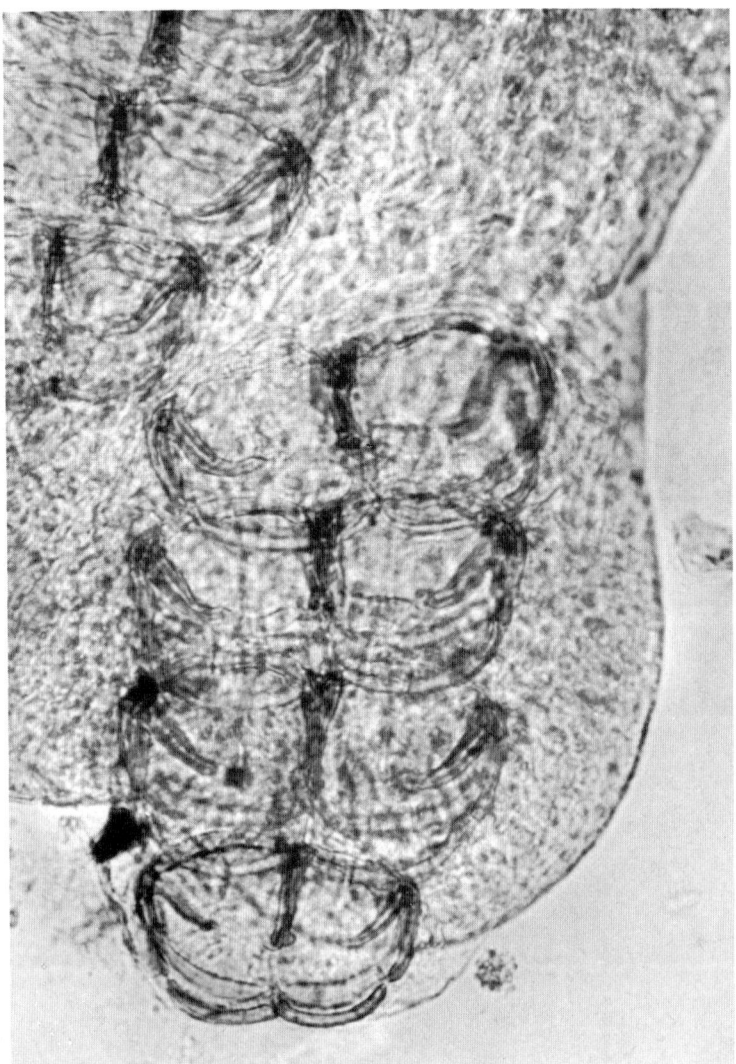

Fig. 21

Figs. 19–21. Irregularities in the clamps of *Diplozoon homoion gracile* R.-Kl.

Treatment of Salmon Lice
(Lepeophtheirus salmonis Krøyer) with Neguvon

P.O. BRANDAL and E. EGIDIUS[1]

Abstract

Serious epizootics of salmon lice, *Lepeophtheirus salmonis,* have occurred in recent years in Norwegian salmonid sea farms. A bath treatment of 300 ppm Neguvon for 15 to 60 min effectively eliminates the parasite. The treatment is applied using a special floating pen enabling the same solution to be used for several fish populations. The method appears to be 100% effective and there are no reports of fish dying as a result of treatment. Fish should not be harvested within 3 weeks of treatment. In Norway the method is approved by the drug authorities, the Neguvon is to be used under veterinary prescription.

---

[1] Institute of Marine Research, C. Sundts gt. 37, 5000 Bergen, Norway

Subject Index

Abramis brama 85, 244
Achromobacter 138
Acinetobacter like bacterium 154
adhesin 221
Aeromonas hydrophila 89, 114, 135, 138, 140, 198, 218
  hydrophila punctata group 90, 137
  liquefaciens 199, 206
  punctata 138
  salmonicida 82, 83, 85, 87, 94, 98, 107, 109, 113, 120, 126, 127, 218
    antibodies 109
    carrier 87
    epizootics 83
    infection 85
    occurrence 82, 87
    vaccine 87, 107, 113
    virulence 104, 223
  subspecies anaerogenes 139
agglutination 62, 99, 108, 109, 115, 151, 218
aggression, fish 225
amoeba 174
Anguilla anguilla 39, 225
antibiotics 78, 90, 107, 128, 150, 155, 224
antibodies 8, 28, 62, 109, 114, 115, 187
antibody like-non virus induced-protein (6 S) 29
antivibriosis vaccine 45
Apicomplexa 165
aquaculture 198
Argulus sp. 135
atlantic salmon 85, 98, 107, 154, 165
  cells (AS) 186

bacteria 37–144, 147, 154, 198, 199, 206, 212, 218, 224
bacterial diseases 37–144, 147, 154, 206, 212, 218, 224
  bacterial kidney disease 147, 199
  carp erythrodermatitis 120, 126, 127, 137
  furunculosis 82, 87, 94, 98, 107, 113
  vibriosis 39, 45, 53, 60, 69, 75
bacterial kidney disease (BKD) 147, 199

bacterial hemorrhagic septicemia 206
bacterial stress 224
bacterin 54, 61
BF-cells 30
blood sporozoa 157
bluegill 228
brown trout 3

calcification 193
calcinosis 199
Carassius auratus 82
carp 23, 28, 82, 157, 187, 228
carp dropsy 127
carp erythrodermatitis 82, 120, 123, 125–127, 137
  histopathology 127–136
  immunization 125
  mode of infection 137
  pathogenesis 120
  pathology 123, 127
  toxin 120–125
CE = carp erythrodermatitis
cell culture adapted virus 29
cell-mediated immunity 115
c-hemoprotozoon 161
chinook salmon 165
Chlamydiae 229
chlorination 79
chronic dropsy syndrome of carp 127
CHSE-cells 187
Citrobacter 138
cohabiting experiment 85
coho salmon 57, 65, 212
Colibacteria 199
corticosteroid 85, 88
cortisol 225
corynebacterium 147–149
  diphtheriae 147
  pyogenes 149
CPE 8, 188
Cyprinus carpio 28, 82, 228
Cytophaga sp. 76
cytotoxicity 104

Dermocystidium 165
Dicentrarchus labrax 229
Diplozoon 235

eel 39, 70, 225
egtvedvirus (VHS-virus) 3, 8, 18, 185, 186
  antigen detection 8, 185, 186
  control 5
  defence 18
  epizootiology 4
  infection 8, 20
  penetration 18
  pike 8
  reservoir 4, 5
  serotypes 3
  susceptibility of trout 3
emerging problems 145–181
endogenous peroxidase 187
Enterobacteriaceae 138
EPC-cells 23
epitheliocystis 228
erythrodermatitis of carp = carp erythrodermatitis
Erwinia 199
Esox lucius 8
exophthalmus 199

FHM-cells 4, 8, 83
fish α-globulin 105
flatfish 75
Flavobacterium 138, 199, 212
flavobacteriosis 212
fluorescent antibody technique 4, 8, 120, 185, 186
furuncle 95, 100
furunculosis 82, 87, 94, 98, 107, 113
  antibodies 114
  immunity 115
  occurrence 82, 87
  vaccination 87, 107, 113

gangliosides 73
general adaption syndrome (GAS) 225
gill biopsy 24
  structure 18

haematopoietic organs 95
haemogregarina 75
haemolysis 70, 99, 101, 154
haemolytic activity 100
  toxin 69
Haplosporidia 174
haplosporosomes 176, 178
hemagglutination 219
hemorrhages 8, 134, 199, 213

Herpesvirus scophthalmi 75
Hoferellus cyprini 163
horseradishperoxidase 187, 188
hyperosmotic infiltration (HI) 30, 46, 61, 107, 113, 207

ichthyotoxic material 98
IHN-virus = infectious hematopoietic necrosis virus
immunity 49, 114
immunization against bacteria 45, 50, 53, 60, 107, 113, 125, 206
  against virus 23, 28, 29
immunodiffusion 33, 104
immunoelectrophoresis 103
immunofluorescence 4, 8, 120, 185, 186
immunoperoxidase technique 186–192
inactivation of virus 30
infectious hematopoietic necrosis virus (IHN) 8
infectious pancreatic necrosis virus (IPN) 29–36, 186
  antigen detection 186, 190
  attenuated virus 34
  avirulent virus 29
  disrupted virus 32, 33
  inactivation 30
  serotype 30
  6-S-resistance 29
IPN-virus = infectious pancreatic necrosis virus

kidney disease 147, 157, 199
Klebsiella 138

Lactobacillus 151
latex agglutination 115
lecithinase activity 100
Lepeophtheirus salmonis 248
Lepomis macrochirus 228
Leuciscus leuciscus 85
leucocytes 99, 185, 225
leucocyte migration inhibition 115, 116
lice of salmon 248

macrophages 99, 167
Marteilia refringens 174, 176, 178
Marteilia sydneyi 176
Mesentericus subtilis 199
Minchinia nelsoni 179
mirror carp 137
Moraxella 138
Morone americanus 228
Morone mullet 228
Mugilidae 228
Myxosporidium 157, 180

Subject Index

Neisseriacea 154
Neodiplozoon barbi 235
nephrocalcinosis of trout 193–205
neutralization of virus 4, 8, 26, 28, 29
neutralizing antibodies 28, 29

Oncorhynchus sp. 147
Oncorhynchus kisutch 45, 212
Oncorhynchus tshawytscha 45, 165
oxygen supersaturation 198

parasitic disease 157, 165, 174
Pasteurella 138
Perca fluviatilis 82, 85
peroxidase 188
Perkinsus 165
PFR = pike fry rhabdovirus
PG-cells 186
phytohaemagglutinin (PHA) 117
pike 5, 8
pike fry rhabdovirus (PFR) 8, 16
pillar cells (PC) 18
plaque neutralization 4, 8
Plesiomonas shigelloides 138
prednisolon acetat 138
proliferative kidney disease (PKD) 174–181, 193
protease 104
proteolytic activity 100, 103
Proteus providencia 138
Proteus rettgeri 224
protozoa 165
Pseudomonas sp. 117, 138
psychrophilic bacterium 154

rainbow trout 3, 8, 18, 30, 54, 60, 94, 99, 147, 174, 193, 198, 206, 212
renal calcification 193
renal sphaerosporosis 158
resistance test 90
Rhabdovirus carpio (RVC) = spring viremia of carp virus (SVCV)
rickettsia like Organism 229
Rickettsiella 231
RTG-2-cells 8, 20, 30, 99
Rutilus rutilus 82, 244

Salmo gaidneri 53, 85, 87, 147, 174, 193
Salmo salar 85, 154, 165, 212
Salmo trutta 82, 87, 114
Salvelinus fontinalis 85, 193
Saprolegnia 80
Sarcondia 180

Sarotherodon aureus 224
Scardinius erythrophthalmus 82
Scophthalmus maximus 75
sea bass 229
sea bream 229
Serratia liquefaciens 138
serum inhibition 103
silver carp 224
skin ulcer 128, 154
Solea solea 75
Sparus aurata 229
sphaerosporosis 157
spray vaccination 208
spring viremia of carp (SVC) 23, 28
spring viremia of carp virus (SVCV) 8, 23, 28, 186
Staphilococcae 199
Streptococcus 151
stress 224, 225, 198
superinfection 96
6-S-sensitivity 29

taxonomy of *Diplozoon* 235
Thymallus thymallus 85
Tilapia 228
toxin 69
toxic effects 71
Trichodina 75
trypsin 104

Upeneus mullucensis 229
urolithiasis of trout 198

vaccination 29, 30, 45, 53, 60, 87, 105, 107, 206
Vibrio alginolyticus 138
Vibrio anguillarum 39, 45, 54, 60, 69, 75, 113, 117, 216
  agglutinin 62
  haemolytic toxin 69
  histopathology 41, 75, 76
  neurotoxin 71
  serum antibodies 62
  vaccination 45, 53, 60
Vibrio cholerae 73
Vibrio parahaemolyticus 73
vibriosis 39–81
vibriostatic compound 60
VHS-virus = egtvedvirus
viral antigen detection 185, 186
viral diseases 1–36, 185, 186
viral hemorrhagic septicemia of trout (VHS) 3, 8, 185, 186

virulence factor 98, 104
virus =
  egtvedvirus 3, 8, 18, 185, 186
  Herpesvirus scophthalmi 75
  infectious hematopoietic necrosis virus (IHN) 8
  infectious pancreatic necrosis virus (IPN) 29–36, 186
  pike fry rhabdovirus (PFR) 8, 16
  spring viremia of carp virus (SVCV) 8, 23, 28, 186

virus 23/75 3
virus penetration 18
virus vaccine 29
VPR-neg. anaerogenes 138

white blood cells 185
wintering condition 24

zoospore 165

M. H. A. Keenleyside

Diversity and Adaptation in Fish Behaviour

1979. 67 figures, 15 tables XIII, 208 pages
((Zoophysiology, Volume 11)
ISBN 3-540-09587-X

Contents: Locomotion. – Feeding Behaviour. – Anti-Predator Behaviour. – Selection and Preparation of Spawning Site. – Breeding Behaviour. – Parental Behaviour. – Social Organization. – References. – Systematic Index. – Subject Index.

The book investigates the behaviour of fish and the influence it has on their adaptation to a wide range of marine and freshwater habitats around the world. Topics under discussion are locomotion, feeding, antipredator behaviour the organization of fishes within groups and reproduction. Separate chapters are included on spawning site selection and preparation, breeding and parental care. The book demonstrates the great variability within each of these categories, relating it to the ecological setting of the fish. The first comprehensive work of its kind, this book will be appreciated by ethologists, physiological ecologists, and fisheries biologists.

E. Scholtyseck

Fine Structure of Parasitic Protozoa

An Atlas of Micrographs, Drawings and Diagrams

1979. 186 figures, VIII, 206 pages
ISBN 3-540-09010-X

Contents:
Introduction. – Classification. – Abbreviations. – Fine Structure: Subphylum: Sarcomastigophora. Subphylum: Apicomplexa. Subphylum: Microspora. Parasitic Protists of Uncertain Systematic Position.

This atlas utilizes a new method of analyzing electron micrographs allowing a more definitive characterization of protozoan fine structural organization. An electron micrograph and an analytic drawing are arranged, along with a descrptive text on pages facing each other to facilitate comparison.
Although emphasis is placed on the cytological structure of a selection of parasitic protozoa, functional, developmental and ecological aspects are also stressed.
The literature is brought up to date with new information on the ecology of parasitic protozoa and on fine structural studies of their developmental stages in the natural host, in cultural cells and in defined media. Problems of host specificity as well as those of parasite-host relationships are discussed.
The atlas is designed for students, instructors, and researchers in the fields of protozoology. parasitology, veterinary or human medicine, and makes a valuable tool in the study of fine structure in general.

Springer-Verlag
Berlin
Heidelberg
New York

Two International Journals

Parasitology Research
Zeitschrift für Parasitenkunde
Organ der Deutschen Gesellschaft für Parasitologie

ISSN 0044-3255   Title No. 436

Managing Editors: J. Eckert, Zürich; B. M. Honigberg, Amherst; G. Piekarski, Bonn; W. Wülker, Freiburg i. Br.
in cooperation with a distinguished board of coeditors.

This international journal on parasitology includes articles on:
- General, Biological, Medical, and Veterinary Parasitology
- Protozoology, Helminthology, Entomology
- Morphology (Pathomorphology, Ulrastructure)
- Biochemistry, Physiology (Pathophysiology)
- Parasite Host-Relationships (Immunology, Host Specifity)
- Life History. Ecology, Epidemiology
- Diagnosis, Chemotherapy, and Control of Parasitic Diseases

Marine Biology
International Journal on Life in Oceans and Coastal Waters

ISSN 0025-3162   Title No. 227

Editor in Chief: O. Kinne, Biologische Anstalt Helgoland
in cooperation with a distinguished board of editors.

The Editors invite and will consider for publication original contributions to the following fields of research:
- Biological Oceanography
- Experimental Biology
- Biochemistry, Physiology and Behaviour
- Biosystem Research
- Evolution
- Theoretical Biology Related to the Marine Environment
- Methods

Fields of Interest: Marine Biology, Ecology, Zoology, Botany, Molecular Biology, Biochemistry, Microbiology, Genetics, Morphology and Evolution of Marine Organisms, Technology and Methods Used in Underwater Exploration and Experiments.

Springer International

Please ask for subscription information and your sample copy.
Send your request to your bookseller or directly to:

Springer-Verlag, Promotion Department,
P. O. Box 105 280, D-6900 Heidelberg, FRG

North America: Springer-Verlag, New York Inc., Journal Sales Dept., 44 Hartz Way, Secaucus, NJ 07094, USA